PRAISE FOR
BAD MEN

"An eye-opening perspective on the deep history of men harming women. David Buss explains which types of men are most likely to be perpetrators, and why women can be so devastated by the results. He never explains away this behavior, but instead explains it with the hope that a deeper understanding can help solve these problems and free future generations of women from the ever-present fear that we've been forced to accept as part of life."
— Faye Flam, host of the podcast *Follow the Science*

"Buss tackles numerous difficult, sensitive, and controversial topics in this book, but throughout he uses a balanced, evidence-driven approach. This book will serve as an important springboard for understanding more completely men's (and women's) behavior. It will be of interest to a diverse readership. A provocative and interesting read!"
— Dr. Maryanne Fisher, editor of *The Oxford Handbook of Women and Competition*

"Brilliantly analyzes what has been called the most widespread human rights problem in the world. Authoritative, insightful, and sympathetic, Buss's book is a perfect resource for the #MeToo generation."
— Richard Wrangham, author of *The Goodness Paradox*

"Highly informative and engaging. An insightful, sensitive, and honest examination of the sources of sexual conflict and the ways we might be able to reduce its harms. An essential read."
— Catherine Salmon, PhD, co-author of *Warrior Lovers*

ALSO BY DAVID M. BUSS

The Evolution of Desire: Strategies of Human Mating

The Dangerous Passion: Why Jealousy Is as Necessary as Love and Sex

Evolutionary Psychology: The New Science of the Mind

The Murderer Next Door: Why the Mind Is Designed to Kill

Why Women Have Sex: Understanding Sexual Motivations from Adventure to Revenge (co-authored with Cindy M. Meston, PhD)

BAD MEN

THE HIDDEN ROOTS OF SEXUAL DECEPTION HARASSMENT & ASSAULT

DAVID M. BUSS

ROBINSON

ROBINSON

First published as *When Men Behave Badly* in the USA in 2021 by Little, Brown Spark, an imprint of Little, Brown and Company, a division of Hachette Book Group, Inc.

First published as *Bad Men* in Great Britain in 2021 by Robinson

Copyright © David M. Buss, 2021

1 3 5 7 9 10 8 6 4 2

The moral right of the author has been asserted.

All rights reserved.
No part of this publication may be reproduced, stored in a retrieval system, or transmitted, in any form, or by any means, without the prior permission in writing of the publisher, nor be otherwise circulated in any form of binding or cover other than that in which it is published and without a similar condition including this condition being imposed on the subsequent purchaser.

A CIP catalogue record for this book
is available from the British Library.

ISBN: 978-1-47214-633-5 (hardback)
ISBN: 978-1-47214-634-2 (trade paperback)

Printed and bound in Great Britain by Clays Ltd, Elcograf S.p.A.

Papers used by Robinson are from well-managed forests and other responsible sources

Robinson
An imprint of
Little, Brown Book Group
Carmelite House
50 Victoria Embankment
London EC4Y 0DZ

An Hachette UK Company
www.hachette.co.uk

www.littlebrown.co.uk

*For everyone who has suffered
from sexual conflict*

CONTENTS

Introduction   3

CHAPTER 1:   The Battle of the Sexes   7

CHAPTER 2:   The Mating Market   30

CHAPTER 3:   Struggles Within Mateships   61

CHAPTER 4:   Coping with Relationship Conflict   92

CHAPTER 5:   Intimate Partner Violence   118

CHAPTER 6:   Stalking and Revenge After a Breakup   145

CHAPTER 7:   Sexual Coercion   166

CHAPTER 8:   Defending Against Sexual Coercion   208

CHAPTER 9:   Minding the Sex Gap   248

Acknowledgments   279

Notes   281

Index   315

BAD MEN

INTRODUCTION

THIS BOOK UNCOVERS THE HIDDEN roots of sexual conflict. Manifestations of sexual conflict emerge in workplace sexual harassment and in online dating deception. They show up in violence within romantic relationships and in stalking in the aftermath of breakups. And they surface in sexual assaults by strangers, acquaintances, and the ones who profess to love us. Underlying these diverse manifestations of sexual conflict are predictable dynamics that materialize when the two sexes recurrently differ in their optimal mating strategies — when what's good from a male perspective, for example, differs from what's good from a female perspective. Mating strategy differences set into motion coevolutionary arms races between them. Uncovering the causes of sexual conflict, which stem in large part from evolved sex differences in our underlying sexual psychology, can help us to create cures. My hope is that this knowledge will benefit everyone who has suffered from sexual conflict and who cares about its victims, and that it will ultimately help us to reduce the occurrence of sexual conflict and heal the harms it creates.

Although it is easy when thinking about the "battle of the sexes" to fall into the trap of thinking about sexual conflict in terms of men as a group against women as a group, an evolutionary perspective illuminates why that framing is misleading. Sexual conflict is mostly about individual men and individual women interfering with each other in ways ranging from deception in internet dating to sexual harassment in the workplace to sexual coercion by strangers, dates, and mates.

This book focuses on sex differences and individual differences among men and among women, so some clarification about these terms is warranted. Biologists define sex by the size of the gametes — males are the ones with the small gametes (sperm) and females are the ones with the large gametes (eggs). This differs from the many meanings attached to the term "gender," which include the different cultural meanings, social constructions, and psychological identities attached to males and females. In humans, important sex differences in reproductive biology include the fact that fertilization occurs internally within women, not within men. Women, not men, carry the metabolic costs of pregnancy. Women, not men, have breasts capable of lactation. Men can produce a child with the mere act of sexual intercourse and no further investment. Women require a costly nine months of internal gestation to produce that same child.

These sex differences in reproductive biology have created selection pressure for sex differences in sexual psychology that are often comparable in degree to sex differences in height, weight, upper-body muscle mass, body-fat distribution, testosterone levels, and estrogen production. As we will explore in this book, psychological sex differences show up in mating motivations, such as sex drive and the desire for sexual variety. They show up in the emotions of attraction, lust, arousal, disgust, jealousy, and love. They show up in thought processes, such as sexual fantasies and inferences about other people's sexual interest. In short, there are evolved sex differences in human sexual strategies, and these differences are key causes of conflict between the sexes.

Within each sex, however, there exist large individual differences. Some men and women have a strong desire for no-strings casual sex; others opt for monogamy with their "one and only." Some women and men practice the art of deception in the mating game; others opt for honest courtship. Some people remain sexually faithful; others have affairs whenever the opportunity arises. Some sexually harass co-workers with impunity; others are appalled at workplace misconduct. Because of these profound individual differences within each sex, all statements about sex differences in this book carry the always-necessary qualifier of "on average." I trust that the reader will understand this point and infer this qualifier in each instance throughout the book to free the writing of the technically correct but cumbersome insertion of the repetitive phrase "on average."

An evolutionary lens helps to identify the specific social circumstances that increase and decrease the likelihood of sexual violations, as well as the specific types of men — the subsets of men marked by the Dark Triad of narcissism, psychopathy, and Machiavellianism — in whom a harm-inflicting sexual psychology is strongly activated. Social circumstances influencing sexual conflict include laws regulating sexuality and their enforcement; cultural norms about permissible and impermissible sexual conduct; marriage systems that permit or forbid polygamy; an individual's coalition of kin and friends who can function as "bodyguards"; the different mating strategies being pursued by individuals within the group; defensive strategies available to potential victims; the ratio of women to men in the relevant mating pool; and many others. One goal of this book is to highlight the social and personal circumstances that reduce or amplify sexual conflict in order to prevent victimization and minimize conflict between the sexes. An evolutionary perspective is not the only one that can help to accomplish these goals, but I believe it provides an important and novel set of tools for doing so.

This book also focuses mostly, although not exclusively, on heterosexual women and men. We now know a great deal scientifically about these populations, but there has been less research on conflicts among gay men and among lesbian women, among bisexuals and pansexuals, and among those across the rainbow of sexual identities and orientations. There is a critical need for research to fill this gap.

To a nontrivial degree, we are all locked in the interior of our own minds. We use introspection about our minds to infer the inner landscape of other minds. Because the sexual minds of men and women differ in some respects, these inferences lead to predictable mind-reading errors. It is my hope that the information in this book helps men and women to obtain a deeper understanding of each other's sexual psychology.

Sexuality is a domain that is highly personal for most people. Material in this book covers difficult issues that have personal relevance for many, and which may be disturbing to readers. They are deeply disturbing to me. These include sexual deception in the mating market; sexual harassment in the workplace; sexual assault by strangers, acquaintances, and romantic partners; intimate partner violence and stalking; suicidal ideation; and murder. If you have been a victim of sexual conflict, know someone who has been a victim, or would like to prevent yourself and others from becoming victims, I hope that this book will help. And I hope that the insights revealed will lead to greater compassion for victims and help curb the darker sides that reside within us.

CHAPTER 1

THE BATTLE OF THE SEXES

> It sometimes seems men are the enemy, the oppressors, or at the very least an alien and incomprehensible species.
>
> — Carol Cassell, *Swept Away*

THE "BATTLE OF THE SEXES," it seems, is reaching a feverish pitch. The #MeToo movement, the rage of the incels (involuntarily celibate men), toxic masculinity, and the rise of the "manosphere" are all key cultural signs. Although the labels and cultural movements are new, the underlying conflicts they reveal are ancient. The battle of the sexes may be traced back in time through human evolutionary history, through primate and mammalian evolutionary history, and even to the origin of sexual reproduction itself 1.3 billion years ago.

Every person alive struggles with sexual conflict. Most of us see only the tip of the iceberg — dating deception, a politician's unsavory sexual grab, the slow crumbling of a once-happy marriage, a romantic breakup that turns nasty. These flash points make big news when their players are prominent — consider the sexual scandals of Bill Cosby (imprisoned), Al Franken (resigned from Congress), Harvey Weinstein (imprisoned), Bill O'Reilly (fired or resigned), Matt Lauer (fired or resigned), Charlie Rose (fired or resigned), or Jeffrey Epstein (imprisoned followed by death due to suicide or murder).

Social scientists struggle to explain why women and men seem so much at odds with each other. Popular explanations blame patriarchy, masculine hegemony, and toxic masculinity. Men, some scholars argue, maintain a vise grip on power and resources, put up glass-ceiling barriers, and exclude women from the old-boys' club. Manosphere bloggers, on the other hand, blame women who seek sex with "alpha chads" (high-status males) and exploit lower-status men who are "betas" for their investment.[1]

There is some truth to each of these contrasting accounts, but also ways in which they lack explanatory depth. Masculinity does indeed have toxic elements. It is no secret that men have a virtual monopoly on sexual harassment, sexual assault, and crass sexual objectification of women. Patriarchal institutions such as laws that give husbands control over their spouses' sexuality, for example, are still on the books in some countries and have lingering pernicious effects in others. On the other hand, it is also true that women, as many in the manosphere claim, tend to be attracted to men who have power, status, influence, and resources.[2] Some women spurn or ignore men lacking high-status attributes; some men feel invisible when it comes to women. Missing from these manosphere accounts, however, is that women's mate preferences are enormously complex and include qualities such as honesty, intelligence, dependability, moral character, sense of humor, and many more. What social scientists, patriarchy theorists, and manosphere bloggers fail to see are the hidden roots of sexual conflict, the underlying causes of the never-ending battle of the sexes. They fail to see how ancient sexual conflicts gave rise to the fundamental evolved components of our sexual psychology. And they fail to see how the ancient sexual psychology we house in our large brains misfires in the weird modern world in which we live — a world filled with novel cultural products such as internet dating, pornography, video game–addicted men, and gender-integrated workplaces. To understand why the sexes get into so much conflict, we must take a few steps back from its modern manifestations. We must realize that the roots of sexual conflict are not uniquely human, although their manifestations in our species take unique forms.

SEXUAL CONFLICT WRAPPED IN SILK

Consider a spider from the family *Pisaura mirabilis*.[3] To attract a female, the male must capture an insect to offer as a nuptial gift, an arduous task at which it fails as often as it succeeds. Females sometimes accept offers and copulate with the gift-giving males while consuming the tasty dinner. But sometimes they seize the gift and flee before copulation, leaving sexually frustrated males in their wake. Fleeing females seem to have won the first salvo of this battle — a free meal at the male's expense. Male spiders, however, have evolved two counterstrategies. First, they are sometimes able to cling to the feast and then go limp; they feign death as the female drags her tasty meal to a private place for consumption. When she settles down to her dinner, the male springs back to life and copulates with her while she eats. Sometimes the male succeeds, but not always.

Success comes more often to males who use a second strategy — gift wrapping the meal in a silk package. But if a male spider fails to find tasty food to offer, he will sometimes wrap worthless trash in silk, such as a clod of dirt or an inedible seed. Silk conceals the packet's contents. The silk threads also render the package easier for the males to clasp, preventing female attempts to grab and run. Attractive wrapping lures the female, and the male can copulate while she's busy unwrapping the deceptively alluring bundle.

If she discovers the ruse, finding trash instead of a meal inside, she abruptly kicks the male off her, ending the encounter before he finishes copulating. Moreover, females use scent to detect whether a legitimate meal resides within the wrapping, and they avoid gifts lacking the right scent. However, males sometimes wrap the dregs from a leftover meal they have consumed, leaving a trace scent of food to fool the females. Females, in turn, have evolved to rapidly detect package quality and will sometimes reject offerings perceived as deficient. Female spiders, it turns out, are very picky about whom they mate with. Who knew?

Humans are not spiders, but we are not exempt. The *Pisaura mirabilis* spider example is merely one among thousands, but it illustrates a critical explanatory principle — the principle of *sexual conflict coevolution*. For every tactic one sex evolves to exploit the other, there exists at least one coevolved defense in the other.

Within-species sexual conflict arms races are analogous to arms races between predators and prey. Each increment in speed and agility of prey favors coevolved increments in speed and agility of predators. Gazelles and cheetahs, for example, are locked in perpetual coevolutionary arms races of defenses, offenses, counterdefenses, and counteroffenses. These arms races are *multidimensional*, involving not just speed, but also vigilance, visual and auditory attention, habitat selection, and even mind-reading capacities that permit elements of surprise attack, defensive anticipation of surprise attack, and preemptive evasion. Similarly, adaptations in women to avoid subpar males or to require extensive courtship displays before consenting to sex have created selection pressures on men to circumvent these barriers. Defensive adaptations to deflect sexual advances are countered by sexual persistence adaptations. As we will see in Chapters 7 and 8 when we examine human sexual predators and their female victims, the predator-prey analogy is disturbingly on point.

Sexual conflict theory — the idea that traits beneficial to the reproductive success of individuals of one sex can damage individuals of the other sex, resulting in coevolutionary arms races of offenses and defenses — has been applied almost exclusively to understanding sexual interactions among insects. Male reproductive organs of some species, such as water striders, possess penile spines that damage the female reproductive tract during mating. Black widow spiders cannibalize their mates after copulation if they fail to flee fast enough, using male bodies as meals for their offspring. Male fruit flies inseminate proteins upon ejaculation that induce female ovulation, reduce her sexual receptivity to other males, compromise her immunity, and shorten her life span.

The incel mantra about women's rejection of beta males echoes the sexual frustration of spider males who lose in their mating efforts by failing to wrap their gifts in an attractive-enough package. As a researcher on human mating strategies, I am frequently asked by men, perhaps more than any other question, something in the form of: "I was so nice to this girl. I gave up my weekend to help her move into her apartment. I bought her a housewarming gift. I devoted so much time and energy to helping her out, hoping she would like me. And then she puts me in the 'friend zone' and ends up sleeping with this guy who's a real asshole. Why? I thought women liked nice guys!"

To take another parallel, modern women's exasperation about unwanted sexual attention — stares that last too long, incidental touches that brush up against a breast or butt, persistent advances — from men they deem unworthy echoes the highly discriminating mate choice of their distant female spider cousins. Studies of online dating sites have discovered that most men find many women to be attractive, showing a roughly normal bell-shaped distribution. In contrast, most women find the average man to be below threshold, showing attraction mostly to men in the top 20 percent.

A female friend of mine — a successful, intelligent, and attractive academic — who tried an online dating site illustrates this gender asymmetry. Within two weeks of signing up and posting her photo and self-description, she had received messages from more than five hundred interested men. Hour after hour over many days she scrutinized each one, assessing his character, his intentions, his way with words, his photos. At the end of this arduous process, she selected only one of the five hundred and sent a reply. After a brief coffee date with him, she concluded that he did not exceed threshold. Back to square one. In contrast, studies of online dating sites show that men send messages to women in bulk, bombarding many with the same cut-and-pasted lines. On Tinder, for example, many men "swipe right" on dozens or hundreds of profiles each day, hoping that one or a few might respond in kind. In contrast, one woman who used Tinder exclusively for short-term mating told me that she swiped right on less than 0.8 percent of men she saw on the site and hooked up with only 0.6 percent of those, resulting in a minuscule 0.005 percent of men she saw on the dating app.

Female choosiness — deciding who qualifies for interaction, relationship escalation, and sexual access — is a good candidate for the first principle of mating, and this book will explain why. But it is precisely this female selectivity that creates one key form of sexual conflict, sometimes expressed as resentment from men who fall below threshold. To some men, women do seem wicked when they reject them. But to delve deeper into the hidden roots of modern manifestations of the battle, we must backtrack to ask the critical question: Why would sexual conflict exist at all?

THE MYSTERY OF SEXUAL CONFLICT

Men and women need each other. Cooperation is a cardinal feature of successful reproduction. We fall in love, mutually choose each other, consent to sex, and sometimes commit to a lover over the long run. If we produce children, each partner has an equal evolutionary stake in their welfare. Given the inherent adaptive benefits of cooperation, sexual conflict becomes a mystery that requires explanation.

Consider the simplest case — a difference between the male and female perspectives on the optimum amount of time after an initial meeting before initiating sex. Women often require more time than men to accurately assess the other's mate value. Qualities that enter into the calculus of a man's mate value — his attentiveness, status trajectory, dependability, health, sense of humor, existing commitments, family ties, and genetic quality — require more than a glance to accurately evaluate. Physical attractiveness, an important component of mate value for both sexes, can be appraised in milliseconds — but this typically makes up a smaller proportion of a man's mate value than a woman's. Moreover, sexual mistakes are typically costlier to women than to men. Women risk a poorly timed or unwanted pregnancy, a higher probability of acquiring a sexually transmitted infection, and more damage to their sexual reputation. Yes, sexual double standards still exist, even in the most sexually egalitarian cultures, such as Sweden or Norway. Based solely on these sexual asymmetries, women and men have diverged in the optimum amount of time they prefer to elapse before seeking sex (see below).

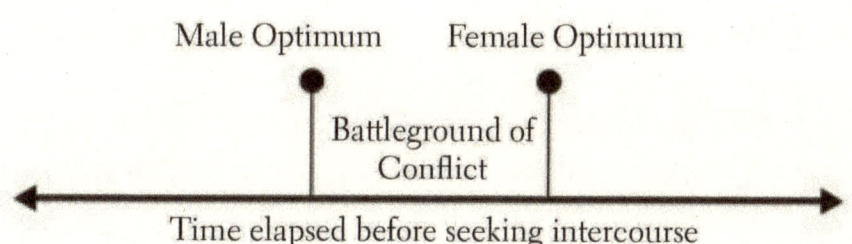

Wherever there exist consistently discrepant sexual optima, sexual conflict adaptations will evolve in each sex to influence or manipulate the other to be closer to its own optimum. One scientist analogized this to having two pairs of hands on the same steering wheel of a car, each having a somewhat different destination, each trying to turn the wheel toward its own destination but being forced to contend with pulls from the other.[4]

Some of the tactics taught in men's pickup-artist-training "boot camps" are specifically geared toward reducing time to sex. For example, one school of pickup-artist thought recommends taking the woman to seven different locations within the span of a single evening, giving her the psychological illusion that more time has passed than actually has. Anecdotal evidence suggests that women are savvy about these coached tactics, and some merely go along before departing with a smile and a goodbye.

Now multiply this one instance of a conflicting optimum by many more:

- The level of emotional intimacy preferred before sex
- Deceptive signals of mate value, with men feigning cues of higher status
- A woman seeking a job while her potential employer is seeking a sexual opportunity
- Disagreements about the frequency of sex within an ongoing relationship
- Conflict about sex with people outside the relationship
- Bank accounts or credit cards concealed from one's partner
- Whether to break up or stay together
- Whether sex will continue in the aftermath of a breakup

Differing optima abound. Sexual conflicts begin when a man and woman first meet on the mating market or online. But they do not end after a relationship is formed. Nor do they end after a romantic breakup. This book will uncover the key sexual conflicts at each of these three temporal phases. We must first explain why conflicting optima exist at all within a supremely cooperative species such as our own, in which women and men need each other for successful reproduction.

THE BATTLE OVER WOMEN'S BODIES

A starting point for understanding sexual conflict is the realization that there exists a battle over women's bodies. Many evolutionary biologists believe that it all starts with the sperm and egg. Men produce millions of sperm, replenishable every hour, that are stripped-down packets of DNA with a tail designed for swimming speed. Women have merely a few hundred viable eggs, large and loaded with nutrients, that they ovulate over their reproductive years and cannot be replenished.

Each egg has a thick external layer, a cortex through which tiny sperm must drill to successfully fertilize the egg. With rare exceptions, only one of the dozens of competing sperm will succeed. Female selectivity occurs even at this level, called cryptic female choice, allowing only the chosen one entry before shutting down shop to all the others — an adaptation to prevent polyspermy. From the moment of conception, when the one tiny sperm joins the nutrient-rich egg, women are already contributing much more than the man.

The asymmetry in investment does not end there. It is the woman who incubates the fertilized egg within her body. It is the woman who transfers calories from her body through the placenta to the developing embryo. It is the woman who suffers through pregnancy sickness, an adaptation to avoid and expel foods containing toxins and pathogens dangerous to the fetus. It is the woman whose center of gravity is shifted increasingly forward, putting more torque and strain on her spinal column, impairing her mobility and rendering her more vulnerable to falling and injury. It is the woman who bears the burden of nine full months of pregnancy, an astonishingly long investment compared to most mammals. And it is the woman who nourishes the infant first with colostrum, which contains key ingredients that help the newborn develop a strong immune system, and then with breast milk — lactation that continues typically for years after birth — to shift metabolic energy from her body to the child's.

A woman's body, in short, through heavy investment that is mandatory in human reproductive biology, has become an extraordinarily valuable reproductive resource for men. A body that incorporates the man's small packet of DNA, adds her own DNA, nurtures the conceptus from the start, protects and feeds it for nine months, immunizes it after birth, and continues to contribute calories for years is an astonishingly bountiful reproductive asset. Feminist theorists correctly observe that men objectify women's bodies and that sexual objectification inflicts costs on women ranging from eating disorders to hiring and promotion discrimination. What is needed, however, is to explain why many men have a psychological proclivity to objectify women. The momentous investment asymmetry in human reproductive biology, together with the sexual psychology that evolved to accompany that biology, provides crucial explanatory pieces.

Men's bodies, in the cold calculus of reproduction, are more expendable. The sexual asymmetry may have started with the fact that sperm are cheap and eggs are expensive, but it did not end there. Indeed, the sexual imbalance, although present in all mammals, became exacerbated in humans due to the prolonged magnitude and duration of women's investment. The sexual asymmetries in human reproductive biology favored the evolution of asymmetries in sexual psychology, which in turn provide the keys to understanding why women's bodies have become one of the main battlegrounds for sexual conflict.

THE STRANGE CASE OF THE TIWI OF NORTHERN AUSTRALIA—WOMEN AS ECONOMIC CURRENCY

A stark illustration of the battle over the female body is captured by a superb ethnography of the Tiwi—an aboriginal group residing on the islands of Melville and Bathurst about twenty-five miles off the coast of northern Australia.[5] They arrived by canoe or boat from southern Asia at least twenty thousand years ago, and likely forty thousand years ago, before cultural innovations such as the bow and arrow, pottery, and agriculture were invented. The word "Tiwi" means "the people" because the Tiwi divided the world into the true people and everyone else.

Although all cultures have rules or laws about marriage, Tiwi custom was highly unusual—all women, regardless of age or circumstance, from the youngest female infant to the oldest grandmother, were required to be married at all times. The first method of marriage was *infant bestowal.* As soon as a female was born, the husband of the woman who bore the child had the right to bestow her on another man in marriage. Bestowal was not random, and those doing the bestowing had to take many factors into account. Sometimes a man would bestow his daughter on an ally or a potential ally—one who could reciprocate by bestowing a wife on him. Sometimes the bestowal would be repayment, returning an obligation to a man who had previously given him a wife. Often, men scrutinized the field of contenders, looking for promising men who were good hunters and likely to rise in rank over time and become "big men" who wielded power and influence.

Rarely were the recipients of the bestowal younger than forty, since older men had higher status, which resulted in large age discrepancies between husband and wife. The female child remained with her natal family until puberty, roughly age fourteen, at which time she moved to her husband's household and began her wifely duties—gathering, cooking, household chores, and procreation.

Men in the group attended closely to the men on whom other men bestowed brides. A Tiwi man might remain entirely mateless until age forty before acquiring his first wife. The other men, inferring that the recipient must be rapidly rising in status and influence, might then bestow their infant daughter on him — a form of mate-choice copying. A bachelor might suddenly find himself with several brides. In the Tiwi past, it was not uncommon for a man to have a dozen or more wives in his lifetime, although not necessarily at the same time, since he had to wait until each reached reproductive age to join his household. One man named Turimpi had more than twenty wives. This mating system left many men with zero wives. This level of extreme polygyny, or *reproductive skew*, as it is called in evolutionary biology, is rare for traditional hunter-gatherer groups. In most of these foraging groups, a highly polygynous "big man" rarely had more than three or four wives.

The large age discrepancies among the Tiwi produced a problem — the old men would die, leaving behind widows. But since all Tiwi women must be married at all times, according to Tiwi law, a second marriage custom kicked in called *widow remarriage*. This proved to be the provision by which some men acquired their first wife. Although the more desirable young females went to existing allies or rapidly rising men, not so the older widows. The final decision of whom a widow was to remarry turned out to be too complex for the anthropologists to determine definitively, but the husband's brothers, other kin, and the widow herself seemed to have a say. Widows were valued and a man with a former widow as a wife attained some level of prestige, especially because many men had no wives at all.

Because the Tiwi marriage customs resulted in large age discrepancies between husband and wife, and women typically were attracted to men just a few years older rather than decades older, problems arose. Young wives found men closer to their own age, rather than their geriatric husbands, most appealing. Given the many unmarried Tiwi bachelors combined with literally zero single women on the mating market, mateless men's attraction to young married women often proved intense. As in all cultures, clandestine trysts occurred. The old husbands, of course, were not oblivious. They were constantly suspicious of young men who seemed a bit too close for comfort. Moreover, older wives sometimes served as spies, revealing to their husbands the treachery of the younger wives. Infidelities were often outed.

According to Tiwi custom, an infidelity was a grave offense committed by the young male interloper against the rightful husband. They had a procedure for handling the sexual conflict — a duel of sorts, although the aggrieved husband had a huge advantage. After an accusation against the "mate poacher," the entire group gathered in a circle. Within the circle stood the accused. A few dozen paces away stood the old man, dressed in ceremonial garb and face paint, armed with both ceremonial and hunting spears. The old man commenced to harangue the accused, citing a litany of all the benefits that he, his relatives, and his allies had bestowed upon the young man since his birth. He contrasted this avalanche of goodwill with the scurrilous betrayal by the young man. After twenty minutes, the old man would cease talking, put down his ceremonial spears, and start launching the hunting spears at the young man.

This was no idle game. Hunting spears can be lethal. But the young man, being spry and agile, could often dodge the javelins, although he had to remain more or less in the same location while spear after spear was hurled. Running away was not an option. If he dodged them for too long, the older man's allies would join him, until soon three or four men were launching spears. Dodging one spear thrown by an old man past his physical prime from thirty or forty paces may be easy; dodging a handful thrown by several men simultaneously is nearly impossible. The best strategy for the accused was to stop dodging and attempt to maneuver his body to receive a spear in the arm or thigh, causing a great gush of blood to a nonlethal body part that would heal quickly. This would salvage the old man's honor and allow the young man to remain in the group without incurring further punishment.

Lacking a cash economy, Tiwi culture treated women as the key currency. To attain a position of prestige and influence, it was imperative that a Tiwi man have many wives. He had to produce many daughters himself to bestow on other men, thereby encumbering them with reciprocal obligations. He had to have the acuity to identify which men were the "best investments," those who were most likely to return dividends of bestowals. And he had to live long enough to reap these dividends, for the many years it took for each bride to reach reproductive age.

The Tiwi were cultural outliers in many respects: in their lack of many modern cultural products; the extremity of their polygyny; their cultural requirement that all females, regardless of age, must be married at all times; and their unique method of handling infidelity, which they referred to as the treasonous seduction of their wives. It may be the only culture on earth that had a total absence of single women and an out-of-marriage birth rate of zero. But the Tiwi share with most cultures the central notion that women embody the most important resource over which men compete. They share with other cultures that conflicts between men often center on competition for women. They share with other cultures that status is a major means by which men attract or acquire women as mates, and reciprocally, that success in acquiring mates bestows status. They share with many other cultures the stance that sex with another man's wife is a crime against the husband, and that the guilty mate poacher warrants punishment.

The Tiwi also illustrate several key forms of conflict between individual women and individual men over sexuality and mating. A husband typically wants his wife to remain faithful, but she is sometimes attracted to other men and sometimes acts on that attraction. Sexual infidelity is common among the Tiwi, as it is among cultures the world over. Sexual jealousy is a key cause of conflict within mateships — a universal emotion stoked when there is a major threat to a valued relationship. Importantly, sexual conflicts within relationships influence other forms of social conflict, such as between an interloper and the husband, among women, and among the larger coalitional alliances of men and women. Sexual conflict, in short, permeates many deep strands of human social life.

THE COSTS OF SEXUAL CONFLICT

Conflict is costly. It benefits neither opponent to do battle. The harms are typically greatest for those who lose the struggle. In the case of some forms of sexual conflict, if the woman loses, her brain and body are manipulated to be under the man's control, doing his bidding, serving his selfish interests rather than her own. Her harms do not end there. To use an analogy, if someone steals from you, your loss is not merely the value of the goods that were stolen. The costs are compounded by all the effort you expended to deter the theft to begin with — locks, alarms, security cameras, guards, Mace, avoidance of dangerous people, steering clear of risky locations, and the psychological vigilance required in an attempt to avoid becoming a victim in the first place. The damage to women, in short, often includes the expenses of her defenses marshaled to prevent being manipulated.

If the man loses in some forms of sexual conflict, he has wasted his time, effort, energy, and resources in a failed attempt to solve a critical adaptive problem — successful mating. He also suffers opportunity costs — the time and effort that could have been allocated toward a more pliable or exploitable target. Even if there is no clear winner and loser, both man and woman have incurred the costs of effort that could have been allocated to other adaptive problems of survival, mating, or investment in children. And win or lose, both have wasted effort devoted to developing and deploying offenses and defenses — the psychological and behavioral tactics required to influence the other to be closer to the actor's own optimum.

In short, there is no adaptive value of sexual conflict per se. Many conflicts and their outcomes are purely maladaptive byproducts for both sexes. If women and men could agree in advance on a compromised middle-ground solution that was perfect for neither but acceptable for both given the circumstances, they could avoid many of these costs. For each offensive adaptation, however, selection favors defensive adaptations in the other, producing a never-ending coevolutionary arms race — an endless cycle of reciprocal adaptations and counteradaptations. Like the Red Queen of Lewis Carroll's classic, each must continue running as fast as possible just to avoid losing ground.

CULTURAL EVOLUTION AND EVOLUTIONARY MISMATCHES

Genetic evolution is a slow process. Cultural evolution is swift. Human sexual psychology evolved gradually over millions of years. These psychological adaptations are products of the past, and they evolved to solve ancient adaptive problems. Yet this primal psychology, so exquisitely functional in times long gone and long forgotten, becomes expressed in novel cultural contexts that ancestral humans never encountered. These discrepancies create mismatches between our evolved psychology and modern cultural environments.

To illustrate, consider male sexual jealousy. Abundant evidence reveals that sexual jealousy is a universal human emotion, variable in its expression but present in all cultures.[6] The intensity of sexual jealousy and the frequency with which it is experienced are roughly equal in men and women. Neither sex has a monopoly on this powerful emotion. The psychological nature of jealousy, however, differs between women and men. Although there is considerable overlap, men's jealousy focuses more heavily on the *sexual* aspects of their partner's conduct. It does so for an extremely important functional reason. It evolved to solve the problem of paternity uncertainty. Women are always 100 percent certain that they are the mothers of their offspring. No woman ever wondered, as an infant was emerging from her body, "Is this kid really mine?" Men can never be sure. As people in some cultures say, "Mama's baby, Papa's maybe." It's an asymmetry that stems directly from an asymmetry of human reproductive biology—fertilization occurs internally within the woman's body, not the man's. And unless a woman is under lock and key twenty-four hours a day guarded by a phalanx of eunuchs, another man might have the opportunity to inseminate her.

Consequently, ancestral men recurrently faced an adaptive problem no woman in the history of human evolution has ever faced—investing resources in the mistaken belief that a child has sprung from his own loins and not from those of an interloper. Male sexual jealousy evolved as one solution to this adaptive challenge. Jealousy motivates men to monitor the sexuality of their partners. It motivates men to be vigilant of interested rivals. It motivates them to guard their partners using tactics ranging from vigilance to violence. And it has led to harmful cultural inventions, such as chastity belts, infibulation, clitoridectomy, and medically misguided virginity tests such as looking for bloody bedsheets on the wedding night after consummation or examining the clarity of a female's urine. In modern times, cultural evolution has produced sophisticated spy cams, internet monitoring, and cell phone tracking devices. A recent example comes courtesy of the Saudi Arabian government, which launched Absher, an app that allows men to track the movements of their wives, even alerting them when women try to leave the country to travel (married women in Saudi Arabia are forbidden to exit the country without their husbands' permission). Sexual jealousy, expressed in ancient and culturally modern forms, motivates men to increase the odds that they are the actual genetic fathers. It also minimizes "wasting" their investment on children sired by rival men — a costly endeavor in reproductive currencies.

Then along came a novel cultural invention that changed everything — reliable hormonal birth control pills, which received FDA approval in 1960. Within a few years, millions of women were taking the pill. In the United States, 98 percent of sexually active women have taken a hormonal contraceptive at some point in their lives, and 62 percent of reproductive-age women are currently on the pill. Comparable figures in Nordic countries such as Denmark, Sweden, and Finland hover in the low 40 percent range. The pill's invention and spread had many consequences for human sexuality. It liberated women from unwanted and untimely pregnancies, for example. But how did it influence male sexual jealousy?

Consider this thought experiment. You are interviewing a newlywed man in a study about his sexual attitudes. You explain to him that although sexual jealousy evolved partly to solve the problem of paternity uncertainty, his blushing bride is taking highly effective birth control pills. Moreover, in the unlikely event that she gets pregnant (less than 2 percent chance if used as directed) you can conduct a DNA test to be 100 percent sure the child is your own. Therefore, the main adaptive rationale for the existence of sexual jealousy is entirely absent. Paternity uncertainty is an ancient adaptive problem that is no longer relevant. After you explain this, you ask the man: "Would you be okay with another man having sex with your wife this evening, now that you know that there's no need for sexual jealousy, as long as there is a guarantee that she won't leave you?"

Would most men be okay with other men having passionate

sexual intercourse with their beautiful brides or trying out different sexual positions with them? If this thought experiment doesn't convince you of the answer, our empirical studies might. When we brought men into the physiological lab and asked them to imagine their pill-taking wives or girlfriends having sex with other men, their physiological distress levels spiked — heart rate, sweating, and corrugator contraction (frowning) went through the roof. When we watched the experiment unfold through the one-way mirror, we observed some men starting to vibrate. Their faces reddened. Metaphorical steam came out of their ears.

This example illustrates a few key points. First, cultural inventions such as the birth control pill fundamentally change the ground rules within which human sexuality gets expressed — they alter the cost-benefit calculus of sexual behavior, in this case severing the link between sexual intercourse and conception. Second, some components of our sexual psychology, exquisitely adaptive in the past, may no longer be adaptive in modern cultural contexts. Third, some aspects of our sexual psychology continue to get expressed despite the modern absence of the evolutionary pressures responsible for their origins.

Importantly, culture continues to evolve and does so rapidly. We invent cultural products that activate and satisfy our evolved sexual psychology whether or not the adaptations that comprise that psychology are relevant to solving modern adaptive challenges.

We tend to think of cultural inventions as more or less unmitigated blessings, and many are. Dishwashers save time wasted on washing. Food-delivery services save time on shopping. Video-streaming services provide immediate access to an array of exciting movies and shows. Modern medicine extends our lives. But just as there are genetic coevolutionary arms races, so too are there *cultural arms races.*

Conflict between the sexes gets played out on a cultural battlefield. As users of internet dating sites develop better and more effective profiles to generate more hits, consumers become more sophisticated about sifting through the profiles. Scientific studies themselves accelerate cultural arms races. They inform users about the precise angles of selfie photos to post, clear-cut facial expressions to display, and the specific leisure activities to list. As deception on dating profiles becomes more effective, so too does detection of deception. Like predator-prey arms races, cultural arms races evolve. And like standard evolutionary arms races, cultural arms races over sexual conflict can be persistent, maladaptive, and costly to all players.

Cultural inventions can entirely outmaneuver or undermine traditionally evolved adaptations. The date-rape drug Rohypnol is a prime example. As we will discover in Chapter 8, women have evolved an impressive array of defenses to prevent becoming victims of sexual assault. When Rohypnol is secretly slipped into a woman's drink at a bar, it combines with alcohol to disarm her defenses. It produces partial amnesia. It clouds her memory. It renders her unable to clearly recall the sexual assault, her assailant, or even the events immediately preceding and following the assault. Cultural products do not always disarm evolved defenses this dramatically, but many have the capacity to hijack our evolved psychology in profoundly maladaptive ways.

In this book we will explore how the explosion of novel cultural products — such as pornography, internet dating, and virtual-reality sex — can exacerbate some types of sexual conflict.

THE LEVERS OF POWER: WHAT DETERMINES WHO WINS?

An evolutionary perspective provides a cogent definition of power when it comes to sexual conflict — the degree to which each interacting woman and man exerts influence over a contested resource. When the contest is over the woman's body, what are the key determinants of power? One is the *power of proximity*.[7] In cultures such as the Yanomamo of Brazil, men must roam widely in search of large game and so cannot maintain close proximity to their mates; consequently, women have greater influence to make sexual decisions according to their own interests. At the other extreme, some men insist on knowing where their partner is at all times or even refuse to let their partner leave the house unaccompanied to get groceries. Men who maintain maximal proximity are able to exert maximal power.

A second lever of power is *size and strength*. Most men are larger and stronger than most women, a sex difference especially pronounced in upper-body strength. Whereas men's leg muscle mass exceeds that of women's by 50 percent, men have 75 percent more arm muscle mass and exceed women by 90 percent in total upper-body strength.[8] Not all men use their greater size and strength to influence women in sexual conflicts. But the threat of their use, or even the potential for their use, is a source of power for exerting control over women's bodies in sexual conflicts ranging from mate guarding in marriage to sexual assault by a stranger.

A third lever is the *power of numbers*. In small-group warfare typical of our evolutionary past, the coalition with the largest number almost invariably wins. A war party of ten, no matter how skilled, cannot overpower an opposing coalition of fifty. In sexual conflicts, women with many allies — female friends, male friends, brothers, uncles, and so on — have more power than women with few or no allies. This lever of power explains why the practice of exogamy, present in two-thirds of cultures, in which a woman marries out and migrates to live with her husband and his clan, puts women at a power disadvantage. Women have more sexual power in endogamous cultures, the 34 percent in which they remain with their own kin. Having multiple allies gives women multiple weapons of defense. If a brother is not around on any given day, an uncle, a male friend, a sister, or a female friend can provide backup.

A fourth lever of power is linked to the first principle of mating — *choice*. Mating markets with many interested partners afford more choice than markets that are sparsely populated. The power of choice extends to existing mateships. Our research discovered that most women cultivate *backup mates*, ranging in number from one to five, who function as "mate insurance." Women with their own economic resources have greater power to leave bad relationships or trade up to better ones. Women with small children who are economically dependent on their husbands have less choice. This is undoubtedly one reason why divorce rates are twice as high when the woman's income exceeds the man's rather than vice versa.

This book explores how these and other levers of power play out in sexual conflicts on the mating market, within mateships, and in the aftermath of breakups.

THE DARK TRIAD

Sexual conflict theory with sexually antagonistic arms races can take us only so far in understanding the war of the sexes. We must explain individual differences — why only some men and some women are especially prone to inflict costs on members of the other sex.

Some scholars argue that all men are potential sexual predators, but science does not bear out this dismal premise. Most men do not "corner" women by the copy machine, surreptitiously "ass-grab" when the opportunity arises, brag about sexual assault as part of "locker-room talk," or show up to business meetings in bathrobes. Many men would not dream of harming women in these ways, or risk compromising their reputations, their futures, or their moral standards with inappropriate sexual advances, even if they experience sexual attraction and could get away with it. But some men do, and the qualities of this subset of serial harassers and assailants deserve special focus.

Research has hit upon an important discovery: serial harassers score high on the *Dark Triad* of personality traits — narcissism, Machiavellianism, and psychopathy. A hallmark of narcissism is a strong sense of personal entitlement, and this extends to the sexual sphere. Machiavellianism is marked by a social strategy of manipulation and exploitation. High Machs, as they are called, view other humans as mere tools to be used for instrumental aims and discarded. High scorers on psychopathy are deficient in empathy and indifferent to others' suffering, although they often convey a superficial veneer of charm that fools some women. All three elements of the Dark Triad coalesce to create a strategy of social exploitation, of which sexual exploitation is a key component.

Can women be sexual predators? Our research on the Dark Triad suggests yes, but in somewhat different ways. Women who score high in Dark Triad traits are more likely to engage in mate poaching, luring men away from existing relationships for sexual encounters. High-scoring women are also more likely to use sex as a tactic for getting ahead in the workplace.

These findings reinforce the fact that bare-bones sexual conflict theory, although correct in positing different sexual optima for women and men, does not afford enough explanatory power when considered alone. Strategic individual differences captured by the Dark Triad and other traits are required to explain within-sex variation in sexually exploitative tactics.

The next three chapters reveal how the Dark Triad traits combine with the basics of men's and women's sexually antagonistic psychology and play out in each of the three temporal contexts of mating — on the mating market, after a mateship has formed, and in the aftermath of a breakup.

CHAPTER 2

THE MATING MARKET

> Each of us is descended from innumerable generations of men who lied, cheated, charmed, bullied, or killed their way to sexual intercourse, and from innumerable generations of women who charmed, seduced, lied, or manipulated their way to extracting economic privileges in return for access to their bodies.
> — Paul Seabright, *The War of the Sexes*

CONFLICT ON THE MATING MARKET starts when a woman and a man pursue fundamentally different mating strategies. When a strategy pursued by one interferes with the successful implementation of the strategy pursued by the other, it produces *strategic interference*. If a woman is seeking a brief fling and the man she meets at a bar is looking for a wife for life, the strategies are inherently in conflict. Their desires cannot be simultaneously satisfied. One is bound to be disappointed.

Strategic interference on the mating market takes many forms. It can occur over differences in perceived mate value, as when a man is attracted to a woman he perceives as an 8 (on a scale of 1 to 10), but she believes he's not good enough for her. It can occur on a date when one individual pushes for sex sooner or with less emotional connection than the other requires. It can occur when a woman walks down the street and is subjected to unwanted sexual attention such as lewd leering or catcalls. Men who harass women interfere with a cardinal component of their sexual psychology — female choice. And it can occur in any circumstance in which men try to bypass a woman's freedom to exercise selectivity, be it sexual harassment in the workplace or sexual assault from a stranger.

A key cause of sexual conflict on the mating market is one of the largest psychological sex differences ever documented — the desire for sexual variety.

THE DESIRE FOR SEXUAL VARIETY

How many sex partners would you ideally like to have over the month? Or the next ten years? A massive study by Professor David Schmitt of 16,288 individuals residing in fifty-two nations, located on six continents and thirteen islands, from Argentina to Zimbabwe, furnished the answer.[1] Men said they wanted 1.87 sex partners over the next month; women expressed a preference for only 0.78, a bit less than a full sex partner. Over the next decade, men said they wanted six sex partners on average; women ramped up to two. Schmitt also found some individual and cultural variation. Monogamously minded men wanted only one partner, both over the next month and over their entire lifetimes. At the other end, some men desired hundreds of sex partners, with a few reporting a desire for more than a thousand. In Middle Eastern countries such as Lebanon and Turkey, men wanted a tad more than 2.5 sex partners over the next month, whereas in Oceanian countries such as Australia and New Zealand, men wanted only 1.77 sex partners over the next month. The corresponding numbers for women were 0.88 and 0.82. The magnitude of the sex differences in desire for sexual variety was huge by social science standards, more than twice the effect size of most psychological phenomena.

Perhaps averages can be misleading, so let's consider different statistics. What about the percentage of men and women who wanted more than one sex partner over the next month? Here, the sex differences proved even more striking. In South America, 35 percent of the men, but only 6 percent of the women, wanted more than one sex partner over the next lunar cycle. Even in cultures such as Japan, where levels of sex drive appear to be unusually low, six times as many men (18 percent) as women (2.6 percent) wanted more than one sex partner. The sex differences proved to be culturally universal without a single exception. Biological sex trumped even sexual orientation in desire for sexual variety. Across the entire sample, gay men were fairly similar to heterosexual men in their desire — 29.1 percent and 25.4 percent, respectively, wanted more than one sex partner over the next month. For lesbian and heterosexual women, the numbers were still fairly small — only 5.5 percent and 4.4 percent, respectively.

On top of these striking empirical findings, a mountain of evidence reveals that the sex difference in desire for sexual variety occurs in real-life contexts as well. It shows up in many forms of sexual behavior. Prostitution, for example, is overwhelmingly a male consumer industry — roughly 99 percent of customers are men.

Another behavioral marker of the desire for sexual variety is seeking sex outside one's primary and presumptively monogamous mateship. The famous sexologist Alfred Kinsey found that twice as many men as women had experienced at least one sexual infidelity while married — 50 percent versus 26 percent.[2] Although women have started to close the gap in recent years, all studies show a sex difference in infidelity rates of at least 10 percent, and most show a larger gap than that. Moreover, men who cheat do so with a larger number of sex partners. Men seeking sex on the side apparently are serial philanderers. Women are choosier even in this domain, typically having a single affair. And of those women, 70 percent cite love or emotional connection as the key reason for the affair, a finding that points less to women's desire for sexual variety and more to the mate-switching function of infidelity — a topic we take up in Chapter 3. Men's affairs are more motivated by sex with someone new.

Additional behavioral data come from analyses of online dating sites. One study placed fourteen fake male and female profiles on Tinder and studied responses to them.[3] An astonishing 8,248 men liked the female profiles, compared with a meager 532 women who liked the male profiles. Part of this gender difference can be explained by the fact that more men than women sign up for Tinder accounts. But although men who do sign up swipe right on hundreds of Tinder female profiles, fewer than 1 percent of women reciprocate that liking. It has been reported that roughly 30 percent of men on Tinder are actually married, suggesting that they are looking for casual sex on the side, although a representative from Tinder denies this high figure.[4]

Another window into sex differences came about when the dating website Ashley Madison was hacked. This Canadian-based website overtly advertises for people in committed relationships who want sex on the side. Their slogan is "Life is short. Have an affair." A group of hackers apparently was upset not so much by a site that facilitated discreet cheating, but rather by the site's failure to follow through on its promise to delete personal information after users requested it. The hackers threatened to reveal the names of people who actually used the site unless it was shut down entirely. Ashley Madison refused. The hackers followed through on their threat. Dozens of high-profile married men were outed. The married man Josh Duggar, former head of a conservative Christian lobbying group that focused on family values, was revealed to have two different accounts on Ashley Madison and had payed roughly $1,000 in fees to use those accounts. More astonishing was the discovery that 99 percent of the female profiles turned out to be fake, created by Ashley Madison to give the illusion that an abundance of attractive married women used the site. In reality, although there were 20 million men actively using the cheating site, only 1,492 women, less than 1 percent of the total user base, actively used the site.

Another example of the sex-discrepant desire for sexual variety comes from a study done in Florida. Consider this. How would you respond to a total stranger of the opposite sex who approached you on the street and asked, "Hi, I've been noticing you around town; I find you very attractive; would you go to bed with me tonight?" If you are like 75 percent of the men approached in this study, you would say "Yes!" If you are like 100 percent of the women approached, you would say "No way!"[5] Men were flattered. Of those who declined, some requested a phone number and a rain check or cited a girlfriend or fiancée in town to beg off. Women, in contrast, were taken aback. Most women need a bit more time, information, and emotional involvement before consenting to sex with a stranger. The results of this 1989 study have been disputed by some, but they have now been replicated in Austria, Denmark, and the Netherlands. If the stranger is quite attractive, a few women will consent, but the sex difference remains large.[6] Studies of consenting to sex with strangers converge with studies of men's expressed desires, their seeking sex on the side, their patronage of prostitutes, and their online dating behavior in showcasing one of the largest sex differences yet discovered.

The psychological and behavioral evidence, in short, all points to the same conclusion: that men and women differ profoundly in their desire for sexual variety. Individuals differ, of course, in the strength of this desire — the distributions overlap, and some women exceed some men, just as they do in weight or height. Individuals also differ in whether this desire is expressed in actual mating behavior, such as casual sex, affairs, or patronizing prostitutes. Many men choose not to act on their desires. Some lead lives of quiet desperation. They suffer longings unfulfilled due to moral, religious, or reputational considerations, or simply because they lack opportunity. "A man is only as faithful as his options," according to comedian Chris Rock.[7]

Rock's observation is surely an exaggeration. The movie star Paul Newman was widely regarded as the most attractive man in the world during his acting peak in the 1960s and 1970s. Women threw themselves at him, creating an abundance of sexual opportunities few men ever experience. When asked by a reporter why he never strayed and maintained total fidelity to his wife, actress Joanne Woodward, he replied with a metaphor: "Why eat hamburger out when you have steak at home?" This comment elevated his sexual attractiveness even more and drew a larger avalanche of female attention, but there is no evidence that he ever succumbed to temptation.

Whether the desire lies dormant or alternatively bursts into libidinous expression, it's a key cause of conflict, both internal and external. President Jimmy Carter, a deeply religious Southern Baptist at the time, confessed in an interview: "I've looked on a lot of women with lust. I've committed adultery in my heart many times." As far as we know, though, he never acted on that yearning and remained totally faithful to his wife, Rosalynn Carter. He felt guilty about his feelings, though, since he took the biblical injunction from Matthew 5:28 seriously: "But I tell you that anyone who looks at a woman lustfully has already committed adultery with her in his heart."

In short, the large and profound sex difference in the desire for sexual variety is not something that merely rattles around in men's heads. Many men are burdened by lust for a variety of different women, constant cravings that cannot ever be fully satisfied. Sexual desire sometimes bursts forth into action. It explains why a handsome movie star such as Hugh Grant would engage in sexual activity with a prostitute, despite having Elizabeth Hurley, a gorgeous model and actress, as his then steady girlfriend. It explains why then governor of California Arnold Schwarzenegger would sleep with his housekeeper despite having the attractive TV host and author Maria Shriver as his wife. It explains why the actor Charlie Sheen spent many thousands of dollars visiting high-end prostitutes, despite his ability to attract beautiful girlfriends. It explains the rage of the incels, whose sexual desires remain forever unfulfilled as they watch women they want from the sidelines of the mating market.

ATTRACTIVENESS DISCREPANCIES IN THE SEXUAL MARKETPLACE

Americans strongly believe in equality. It was enshrined by the founding fathers in the Declaration of Independence: "We hold these truths to be self-evident, that all men are created equal, that they are endowed by their Creator with certain unalienable Rights, that among these are Life, Liberty, and the pursuit of Happiness." Modern sensibilities, of course, would correctly include women and would read "all people are created equal." There exist two distinct meanings of this key phrase, and confusion between the two has caused much mischief. The first is that all people have *equal rights*, a principle that has expanded legally to include people of all religions, races, colors, creeds, sexual orientations, genders, gender identities, ages, disabilities, and so on. The second is that all people are created *equal in talent*. They are not. It seems supremely undemocratic, and mentioning it risks being accused of being morally repugnant, but the fact is that people differ profoundly in how desirable or valued they are on the mating market.

Differences in desirability create havoc in at least two fundamental ways. The first centers around *misperceptions*. A man who is a 6 but thinks he's an 8 will be utterly irritated when the woman who's an 8 whom he's trying to chat up rejects his advances. Although women and men both can err in their self-perceived mate value, research shows that men are more likely than women to be overconfident in a variety of domains. Men experience higher self-esteem than women — a sex difference that emerges at puberty. Men have higher estimates than do women of their own physical attractiveness, for example, a sex difference robustly documented in studies conducted since the 1980s.[8] Consequently, men are more likely than women to err in overestimating their desirability on the mating market.

A second way in which desirability discrepancies create mating conflict centers not on men's misperceptions but rather on the hard, cruel fact mentioned earlier — men view many women as "above threshold" in attractiveness, but women tend to be attracted primarily to men in the top twentieth percentile. This gives women the upper hand in the sexual marketplace. Sex differences in desirability inevitably leave many men burdened by sexual desires that they can never consummate. To modify a classic cliché, "Hell hath no fury like a *man* scorned."

A third conflict produced by desirability discrepancies is more subtle. It combines several ingredients. To men's greater desire for sexual variety and women's attraction to men in the top 20 percent we must add the fact that men are willing to lower their standards for casual sexual encounters when the investments, risks, and costs are low. Whereas men require women to be in the sixty-fifth percentile in intelligence for a marriage partner, for example, they drop it to the fortieth percentile or lower for a casual sex partner.[9] So a man who is an 8 in mate value is perfectly willing to go to bed with a woman who is a 6 if doing so requires minimum investment. Men are willing to date down when it comes to sex.

Now we add a problematic element to the mix — women (and men) try to secure long-term mates at the upper end of their mate-value range, the most desirable that they can aspire to successfully attract. One study of online dating found that both sexes pursue partners who are an astonishing 25 percent more desirable than they are.[10] Hope apparently springs eternal. A woman in this situation, receiving attention from the higher-mate-value man, typically believes that he is within her mate-value range as a long-term partner, although unknown to her he simply might be seeking casual sex.

One colleague captured the conflict that can ensue by expressing frustration after several years of unhappy dating: "Why am I being pestered by guys I don't care about, but the men I'm genuinely attracted to seem to show little interest in me?" I told her that she is an 8 chasing after 10s but being pursued by 6s. It dawned on her that pursuing men just outside of her mate-value range was the source of her misery. Why would it take my intelligent friend so long to come to this realization? Her belief was encouraged by high-mate-value men who gave her cues to long-term mating interest — acting helpful, taking her to nice restaurants, displaying interest in her personal life, finding common interests that they shared. Men interested in casual sex commonly provide misleading long-term cues because they work. Total transparency by a man in his short-term sexual intentions typically fails if the woman is looking for love rather than a casual fling, which brings us to conflicts arising from sexual deception.

SEXUAL TREACHERY

"Men are one long breeding experiment run by women," according to some evolutionary anthropologists.[11] Men have evolved to be fiercely motivated to acquire the resources and status women desire in a mate and to embody the qualities women want, such as kindness, dependability, and physical fitness. But some men try to fake them. Men's magazines such as *Maxim* routinely publish articles such as "Fake Your Way into Her Bed" that provide explicit tips on how to deceive women. Women's magazines such as *Glamour* to defraud the victim, gain access to bank accounts or credit card numbers, or steal his identity for illegal financial gain. Men's sexual psychology, in short, can be hijacked for fraud. Women, too, can be victims of catfishing, although the lure is more likely to be long-term romance rather than sex, and female victims also risk getting fleeced of money. Victims of catfishing often are too embarrassed to go to the police or to reveal to their spouse or friends that they have been scammed, enabling this form of fraud to flourish.

Neither sex has a monopoly on deception. One study found that an astonishing 81 percent of online dating profiles contained at least one lie about a verifiable characteristic such as age, height, or weight.[16] Just as men are more likely to lie about their height, women are more likely to lie about their weight — shaving off roughly fifteen pounds from their actual weight. Both sexes post photos that are less than 100 percent accurate. Older online daters post photos of their younger selves, sometimes by as many as fifteen years. Sometimes a man discovers that the slender young woman he thought he was going to meet turns out to be plump and middle-aged. Women who say they are looking for "light and casual" sometimes infiltrate the man's mating mind until he wakes up one day in a long-term committed mateship and realizes that he can't live without her — a female version of bait and switch.

Cultural coevolutionary defenses have emerged for protection. Some internet dating sites promise to vet profiles, investigating the veracity of users' claims. In turn, scamming websites have popped up that claim to verify age and other information about potential dates at no charge, and then sign people up for hidden charges or subscription pornography sites that are not discovered until the credit card bills arrive.[17] Cybercriminals, in short, can take advantage of the sexual psychology of both men and women for nefarious ends.

SEXUAL OVER-PERCEPTION AND UNDER-PERCEPTION BIASES

Most psychological states such as sexual desire are inherently unobservable. We are confronted with a chaos of probabilistic social cues. Was that woman's smile a signal of sexual interest or was she being merely friendly? Was her incidental touch on my arm accidental or a deliberate indicator of interest? Inferences about sexual receptivity cascade to the imagination; men engage in sexual scenario building. Will my approach be warmly greeted, or will she shrink back in horror? Which tactics might transform her initial indifference into romantic interest? Social ambiguities require sophisticated mind-reading abilities.

Amplifying this challenge is the fact that men's and women's sexual psychology differs profoundly — a conclusion that generates considerable ideological resistance but is overwhelmingly supported by sexual science. Differences create problems of inference. One's own mind, sometimes a good guide to inferring psychological sexual states of same-sex others, can be a terrible guide when it comes to understanding the other sex.

Our research has documented systematic errors in sexual mind reading. For example, men dramatically *underestimate* how upsetting unwanted sexual advances, such as leering and touching, are to women. At the same time, women *overestimate* how upset men would be if they were sexually harassed by women. Both genders err in sexual mind reading, but in opposite directions.

These inferential biases lead to sexual conflict even over whether formal sexual harassment codes should be implemented in the workplace and at universities. Women at the receiving end of unwanted sexual attention incur the costs of deflecting it and risk not being valued for their skills or intelligence. Women not receiving sexual attention sometimes feel resentment that the men with connections are giving attractive women unfair professional advantages — a collateral cost of sexual harassment typically unnoticed. Many senior men don't see what the big deal is; the costs to women are often invisible to them.

Strange as it may seem, men vulnerable to the *sexual over-perception bias* — over-inferring sexual interest when it's not there — actually believe that women invite their sexual advances. Consider what this man wrote in his diary: "I have one question — if she didn't want me to feather her nest, why did she come into the Xerox room?... She knew I was copying stuff in there. I had my jacket off and my sleeves rolled up, revealing the well-defined musculature of my sinewy arms which are always bulging with desire. I know what she wanted. This didn't require a lot of thought."[18] The man? US senator Bob Packwood, whom the Senate Ethics Committee voted unanimously in 1995 to expel from the Senate for sexual misconduct after nearly twenty women came forward and alleged sexual harassment. Packwood resigned the day before the full Senate vote to expel him.

Our speed-dating lab study of the sexual over-perception bias led to several fascinating findings. We had women and men who had never met interact with each other for five minutes and then evaluate the other on their sexual interest in them and report on the level of their own sexual interest. Then interaction partners rotated, chatted with a new person, and did the ratings again. Each person interacted with a total of five members of the other sex.

Our first finding confirmed the sexual over-perception bias — men over-inferred a woman's sexual interest in them compared with women's reports of their actual interest. Not all men, however, are equally vulnerable to the bias. Some proved to be accurate at inferring women's interest or lack thereof. Men who scored high on narcissism and who indicated a preference for short-term mating were exceptionally prone to this bias — an inferential error that presumably promotes many sexual advances, even if many of them are not reciprocated. Narcissistic men apparently think they are hot, even when they're not.

Not all women were equally likely to be victims of the male bias. Rather, women judged to be physically attractive by the experimenters were especially prone to evoke men's sexual over-perception. The irony is that attractive women, because they receive a larger volume of male sexual attention, are precisely the women who, on average, are least likely to reciprocate men's sexual interest.

We also discovered an error in women's inferences — a *sexual under-perception bias*. When asked whether each man with whom they had interacted was sexually interested in them, women consistently judged the men to be less interested than they actually were. Several possibilities might explain this puzzling discovery. Perhaps women's underestimation of men's interest functions to deflect unwanted sexual attention from men by literally not "seeing" it. Perhaps it is another expression of female choosiness, with most men being invisible as viable options in women's mating minds. Another possibility is that men withhold direct expressions of sexual interest to avoid the embarrassment of being rejected and the negative damage to their reputation that follows.

Or perhaps men who are sexually interested intentionally conceal their interest as part of their mating strategy. Explicit displays of sexual interest, especially early on, typically backfire and turn women off.[19] Just as men going to battle suppress overt expressions of fear they truly experience, and men feeling jealous suppress it to avoid being perceived as insecure, men may suppress the overt expression of their sexual desires precisely as a tactic for successfully implementing those desires.

Missed sexual opportunities historically would have been costly for men in the currency of reproductive success, and over-inferring sexual interest when confronted with ambiguous cues would have minimized missed mating prospects. Although this hypothesis is speculative, women's sexual under-perception bias may be adaptive in deflecting unwanted sexual attention, although it may simply reflect the fact that men often conceal their sexual interest. In the evolutionarily novel modern work environment, these mating mechanisms go awry. Men's sexual over-perception proclivity, for example, is a key cause of sexual harassment — a topic we take up in Chapter 7. Now, though, we must turn our attention to another problematic feature of male sexual psychology — a special attraction to women who are sexually exploitable.

WHICH WOMEN DO MEN PERCEIVE AS SEXUALLY EXPLOITABLE?

Cheetahs on the Serengeti Plain of Africa prey on herds of gazelles. They face a critical adaptive challenge — which one to choose from the herd to attempt to attack. They use stealth to sneak up on their prey, trying to remain undetected as long as possible before springing into action. They can accelerate from zero to sixty-two miles per hour in a mere three seconds — an acceleration comparable to one of the ten fastest cars in the world, the Porsche 911 Turbo S. The main weapons of cheetahs include speed and superior agility, for they can also stop astonishingly quickly on the dead run with their massive clawed paws and change direction as they track their elusive prey. Their choice of victim is influenced by both their ability to surprise and the potential prey's physical condition. They tend to go after those who are smaller, slower, weaker, and in ill health — cues to *catchability*.

Gazelles have evolved an antipredator adaptation known as *stotting*.[20] Stotting is a behavior in which the gazelles leap into the air, lifting all four legs simultaneously, land more or less on the same spot, and then bounce-leap on all fours several times. Gazelles stot only when they detect cheetahs. The tactic serves two possible functions. First, it alerts the cheetahs that they have been spotted and communicates that the hungry predators have lost the element of surprise — one of their key weapons. Second, it signals to the cheetahs that the gazelles are in excellent physical condition. It is as though the stotting gazelles are saying: "I am so athletic, so nimble, so fleet of foot, that you won't be able to catch me. You are better off going after more catchable prey." Stotting works. Cheetahs rarely go after gazelles after watching them stot. They choose their victims carefully.

Exploitative resources-acquisition strategies are those that use deception, coercion, threats, intimidation, terrorization, or outright force. In contrast to cooperative win-win strategies, exploitative strategies, when successful, create a win for the exploiter and a loss for the victim. The simplest human example is robbery or theft — resources are extracted from the victim to line the pockets of the thief. Muggers choose victims who seem *muggable*. Researchers videotaped sixty different individuals as they walked down the same block in New York City.[21] These tapes were then shown to fifty-three prison inmates convicted of violent assault and mugging. Inmates showed strong consensus about whom they would victimize. Those they flagged tended to move in an uncoordinated manner, with a stride that was too short or too long for their height. Non-victims, in contrast, displayed a more coordinated walk, a normal confident stride, with coordinated synchronous foot movement and shifts of body weight. The muggable victims, in short, emitted nonverbal cues that indicated ease of victimization.

Sexual exploiters also choose their victims, and we now have some scientific clues about the bases of their selections. In one study, researchers created short video clips of women walking down a city street in Tokyo and showed them to men attending university there.[22] They obtained personality data on the female walkers and their reports about how frequently they had been inappropriately sexually touched in public in the past. Like muggers, normal college men displayed strong consensus about which women they would choose as potential victims. Nonverbal cues of those they chose included walking slowly and having a short stride length. The women whom men chose for unwanted touching also tended to score high on the personality trait of neuroticism, low on extraversion, and high on shyness. Finally, the researchers found some correspondence between potential targets of sexual advances and the women's self-reported frequency of having been sexually approached in their lives. In other words, women who have suffered from inappropriate touching in their everyday lives seem to emit cues inadvertently that potential sexual harassers can detect. These findings, of course, neither excuse the harasser nor warrant attributing any blame to the victim. But they do provide important statistical associations that have potential educational value.

Physically attractive women were also more often chosen as targets of unwanted sexual advances. Physically attractive women are targeted as victims not because they are easier to sexually exploit (there is no evidence that they are), but rather because choosing attractive victims historically yielded greater benefits in reproductive currencies because they tend to be more fertile.[23] They activate men's sexual circuits. Just as attractiveness seems to evoke in men an especially strong sexual over-perception bias, it also makes women more vulnerable to attempts of inappropriate sexual touching — costs inflicted by men that women are forced to deflect. Evolution by selection, amoral in nature and indifferent to suffering, has forged some nasty human adaptations.

THE PSYCHOLOGY OF SEXUAL EXPLOITABILITY

Researchers in my lab wanted to conduct more systematic studies to identify cues to sexual exploitability. We first conceptually identified potential classes of exploitability cues. These included psychological cues, such as women who are *emotionally manipulable*. Low self-esteem in women, for example, is linked to a higher probability of experiencing sexual coercion. Another psychological candidate we postulated was *gullibility*—unusually high levels of naïveté, innocence, or trustfulness. A third was *low cognitive ability*, reflecting poor mind-reading skills, difficulty in processing complex information, and perhaps an inability to discern exploitative intent in seemingly friendly strangers or affable acquaintances.

Another category of psychological cues is those that involve *flirtatiousness*, promiscuity, or permissive sexual attitudes. These inclinations might cause women to put themselves in situations, such as fraternity parties or singles bars, that potential sexual exploiters also frequent. Yet another class of psychological cues is those that indicate *recklessness* or *risk taking*, as reflected in personality traits such as impulsivity and sensation seeking.

Ease of victimization, we thought, might also be conveyed by *cues of incapacitation*. Intoxication, fatigue, or other forms of cognitive impairment could make a woman less able to fend off tactics of sexual exploitation. These states disarm women's natural defenses, making them more vulnerable. Certain social cues, such as being alone or lacking allies in the vicinity, might also make women more defenseless. Being in social gatherings surrounded by strangers, rather than being in the proximity of friends, family members, or other "bodyguards," can render a woman more exposed. Finally, certain physical cues, such as small stature, low muscularity, low energy level, or slow walking speed, might attract exploiters.

To identify the strongest cues to sexual exploitability, our research team secured 110 digital photographs of women who varied on these potential cues. We then asked a panel of seventy-six men to rate each photograph on the degree to which each woman would be *seducible, sexually deceivable,* and *pressurable,* the ease with which she could be manipulated into having sex. We also had the men rate each woman on attractiveness as a casual sex partner and attractiveness as a long-term mate.

Our findings proved fascinating. First, we discovered a sharp distinction between sexual attractiveness and long-term-mate attractiveness. Often, these were negatively correlated. For example, if the woman was perceived to be *intelligent*, she rated high on long-term attractiveness but low on sexual attractiveness. Men found women perceived as *cognitively disadvantaged* to be easier to seduce, deceive, and pressure into having sex.

A quote attributed to the author of *Fear of Flying*, Erica Jong, notes that "you see a lot of smart guys with dumb women, but you hardly ever see a smart woman with a dumb guy."[24] It has become a cliché that men are intimidated by intelligent women or perhaps apprehensive about women who are more intelligent than they are. This effect is often attributed to "the fragile male ego." Our findings suggest a different explanation — intelligent women are more difficult for men to manipulate and exploit. Bright women are better at cross-sex mind reading, more able to anticipate exploitation, and more successful at marshaling preemptive counterstrategies.

A vivid real-life example of the sexual allure of women perceived as cognitively challenged was provided by a recording of the conservative news host Tucker Carlson, commenting on air about one of the contestants in the 2007 Miss Teen USA contest who had been widely mocked for one of her answers during the contest. He makes clear that he sees her as easy prey, saying, "She's so dumb. She's like, she's vulnerable. She's a wounded gazelle, separated from the herd."[25] Carlson says the teen would "probably be a good wife" and goes on to ask his hosts, "Don't you think — I mean if you had a wife that dumb would it be good or bad?" To which they answer "Good." These disturbing comments reveal that men sometimes view women as sexual prey, making the predator-prey analogy disconcertingly on point.

This leads us to another key discovery — men perceive women who seem to be *immature* and *young* to be more sexually exploitable than women who are somewhat older and more emotionally mature. Men view immature women as attractive for casual sex but not for long-term mating. This partially explains why young women — in their teens and early twenties — are much more likely to be sexually victimized. Lacking much experience on the mating market renders these women more vulnerable. More mature women tend to be wiser, especially after acquiring experience on the mating market. This explains why college women are far more likely to be sexually victimized as freshmen than in any subsequent year. One study found that 73 percent of college-attending victims of sexual assault, which included unwanted sexual touching, were freshmen or sophomores.[26]

Being *intoxicated* and *sleepy* proved to be cues to sexual exploitability. Alcohol weakens physical strength, renders the imbiber less vigilant, clouds judgment, and dulls natural defenses. Alcohol stimulates a set of neurotransmitters called endorphins. Endorphins are widely known to reduce pain. One of their other functions, however, is to induce romantic bonding, a key part of the human attachment system. Women have less alcohol dehydrogenase in their stomachs than men. This enzyme is produced as part of the detoxification process of alcohol by the body. It is the key cause of why women become more rapidly intoxicated by alcohol than men, even when they consume less per body weight. Because alcohol stimulates bonding endorphins, women are more likely to misread interactions and relationships with men when intoxicated. They overestimate the likelihood of an emotional bond and a long-term relationship — what Dr. Andy Thomson calls the *Prosecco perception bias*.[27]

The Prosecco perception bias explains why this sometimes leads to disaster. Campus counselors encounter women who talk about becoming intoxicated and having sex with a guy who they think is a potential romantic partner, only to realize when sober that it was a terrible mistake. The Prosecco perception bias, produced by the alcohol, then the endorphin release, and in turn the misfiring of the bonding adaptation, creates sexual conflict. Men's deceptive mating strategies — feigning deeper feelings than they really have in order to get sex — exacerbate sexual clashes of interest. But women's inferential errors illustrate that men do not have a monopoly on mating-motivated biases — errors played out in novel modern environments such as alcohol-fueled hookup cultures on college campuses.

In the small-group living of our evolutionary past, young women typically were surrounded by kin and other social allies who could deter potential sexual predators by their mere presence. In the modern Western world, women flock to college campuses fresh from high school, unprotected by the proximity of caring kin. They are left to grapple with newfound freedoms, evolutionarily novel drugs, an evolutionarily unprecedented hookup culture, and high-testosterone alcohol-fueled men, some of whom have honed their strategies of sexual exploitation.

College students often mix alcohol with other drugs, such as marijuana or ecstasy. Marijuana, although often considered by users a harmless drug, can alter judgment, ease anxiety, and induce drowsiness — effects that increase vulnerability. Anxiety, an unpleasant emotion often considered dysfunctional, is actually supremely adaptive at appropriate levels. It heightens vigilance to potential dangers. The drug ecstasy typically intensifies sensations by flooding the brain with dopamine. Sometimes called "the love pill," ecstasy can heighten feelings of attachment and social bonding. It simultaneously lowers inhibitions, amplifying sexual exploitability cues inadvertently sent to strangers or casual acquaintances.

Men judge *flirtatious* women to be easier to seduce, deceive, or pressure into having sex. If flirtatiousness includes amped-up smiling, touching, and prolonged eye contact, it can trigger men's sexual over-perception bias.

Another exploitability cue proved to be *clothing* — dressing in tight-fitting or revealing outfits. Women have, and of course should have, freedom of clothing choice. Women's manner of dress does not excuse men legally or morally from being guilty of sexual exploitation, although historically it has been misused by defense lawyers for this purpose. Nonetheless, it is important to have a scientific understanding of how sexy clothing affects male brains and their decision making. Although we are far from having complete knowledge, evidence suggests that women who wear sexually provocative clothing activate the amygdala and anterior cingulate cortex in men — brain regions involved in regulating arousal and desire.[28]

Sexual arousal may incline men to exploit women. A study by professors George Loewenstein and Dan Ariely induced sexual arousal in men by having them watch pornographic images. Compared to nonaroused men, sexually aroused men reported more willingness to engage in a variety of morally questionable sexual actions, from deception to date rape.[29] These included: tell a woman that you love her in order to increase the chance that she would have sex with you; encourage your date to drink to increase the chance that she would have sex with you; keep trying to have sex after your date says "no"; and "slip a woman a drug to increase the chance that she would have sex with you."[30]

Not all men perceive women as sexually exploitable, and even if they do, perceiving women as emitting cues to sexual exploitability does not mean that they will act on those perceptions. One study in our lab, led by David Lewis and Cari Goetz, looked at two key predictors. The first was mating strategy — whether the men were short-term oriented, tending to pursue sex unencumbered by a committed relationship, or long-term oriented, pursuing a strategy of high-investment committed mating. The second was the personality variable of agreeableness, a key opposite indicator of psychopathy in the Dark Triad. High scorers on agreeableness are nice, cooperative, and empathic in social interactions (sometimes called Light Triad traits); low scorers tend to be self-centered, lack empathy, and be aggressive.[31] We predicted that lack of empathy — an inability to feel the pain of other people — would facilitate a sexually exploitative strategy. We also looked at relationship status, whether men were single or in a committed romantic relationship.

Seventy-six men who varied on these dimensions evaluated photographs of women and rated them on sexual exploitability. When a man was not enmeshed in a committed relationship that could be jeopardized by exploitative short-term mating, *and* he possessed low levels of agreeableness as well as a greater orientation toward uncommitted sex, he was more likely to pursue an exploitative short-term mating strategy. Other research confirms that the toxic combination of Dark Triad personality traits and short-term mating strategy is hazardous to women because it is linked with a higher likelihood of sexual assault, a topic explored in detail in Chapter 7.[32]

The sexual arms race, however, did not end with the male sexual over-perception bias, nor with men being schematic for women who might become sexually exploitable. Once these features developed in men's minds, they created vulnerabilities for women to manipulate for their own goals.

HOW TO OPEN DOORS WITH A MERE SMILE

When evolution installed in male brains the sexual over-perception bias and the ability to detect sexually vulnerable women, it proved costly to women. Could women turn the tables and exploit the exploiters? To find out, Cari Goetz, Judith Easton, and Cindy Meston brought sixty heterosexual women into the lab.[33] They informed the women that they were considering creating an online dating site at the University of Texas at Austin. They wanted to videotape each woman. Research assistants informed each that she was to create a one-minute-long video of herself, filmed by the researchers, as though she were "interested in using the site to find someone to date." Researchers also collected information about each woman's preferred mating strategy, short-term or long-term, as well as the personality traits of openness to experience and extraversion, and rated their overall physical attractiveness. Subsequently, five research assistants blind to the researchers' predictions rated each woman on the degree to which she signaled cues known to be perceived as exploitable, such as flirtatiousness, attention seeking, recklessness, and a kind of ditziness suggesting low intelligence.

They found that women who tended to pursue a short-term mating strategy appeared to *intentionally* signal sexual exploitability. Physically attractive women and those scoring high in openness to experience were especially apt to signal exploitability cues. The authors concluded that women's cue displays reveal an active signaling strategy of mate attraction, especially strongly for women dispositionally inclined to pursue short-term mating.

In principle, active signaling might help women achieve their mating goals through several means. First, their displays will attract a larger pool of men, giving women a greater range and quality of potential partners to choose from. Second, women can benefit from a short-term mating strategy in a variety of ways. These range from obtaining tangible resources to evoking jealousy in an existing partner in a way that increases his perceptions of her desirability.[34] Active display of exploitability cues increases the success of that strategy.

Exploitability cues, for instance, can evoke displays of resources from interested suitors, who might lavish women with drinks, dinners, or gifts. Women can acquire these resources whether or not they intend to have sex with any particular interested man. As an example, one female professor told me that she actively smiled, flirted, and acted a bit ditzy with her car mechanic when she dropped off her car for servicing. It's unlikely that the mechanic truly believed that he would "get lucky" with her. Nonetheless, if these cues changed the man's internal calculus from a 0 percent to a 2 percent probability of a sexual encounter, he would be willing to give her excellent service on her car. Women high in physical attractiveness seem more apt to parlay their assets using this strategy. The LA rock band the Eagles appear to be on target in their song "Lyin' Eyes" when they opine that city girls do seem to learn early how to open doors with merely a smile.[35]

Actively signaling exploitability cues, however, comes with costs. It can create unwanted sexual attention from men in whom the woman has zero interest. Deflecting unwanted sexual attention can be costly. Women risk resentment and vengeful retaliation from spurned men. Perhaps that is why women ward off men using tactics least likely to enrage them. Women rarely state the truth of the matter: "You are a total loser. I'm way out of your league. I have zero interest in you. Now, buzz off, creep." Instead, they deflect by failing to perceive it — women's sexual under-perception bias. Some women created a Reddit thread about tactics for avoiding unwanted sexual attention. Suggestions included ignoring, playing dumb, dressing in body-concealing clothing such as a chest-flattening sports bra, and walking to a crowded place.[36]

In any single encounter in these sexual struggles, either the woman or the man might triumph. But there's another twist in the plot — some women are sexually attracted to precisely the men who pursue an exploitative mating strategy.

THE "BAD BOY" PARADOX

Men scoring high in Dark Triad traits — narcissism, Machiavellianism, and psychopathy — turn out to be unusually attractive to women. Gregory Carter and colleagues created profiles of men scoring high in Dark Triad traits.[37] These included self-descriptions using items capturing narcissism (a desire for attention, admiration, special favors, and prestige), Machiavellianism (proclivity to manipulate, deceive, flatter, and exploit others), and psychopathy (lack of remorse or concern with morality, lack of sensitivity to other people, cynicism about other people). Women found these profiles more attractive than control profiles lacking high-level Dark Triad attributes, even when physical appearance was identical and both sets of profiles lacked other cues that women are known to find attractive, such as good financial prospects. The "bad boy" paradox is why women would find high-scoring Dark Triad men attractive, even though these men are socially duplicitous, are unfaithful when in relationships, and have no qualms about abandoning women after sex to seek other sexual prey. As one woman quoted in Urban Dictionary noted, "Look at the bad boys, they are so hot, but dangerous which *turns me on* even more."[38]

Several clues suggest an explanation for women's attraction. First, men scoring high in Dark Triad traits are often socially charming. They are smooth operators who talk a good game. When interested, they focus their attention laser-like on a particular woman, making her think she is supremely special. It's an elixir, especially for women accustomed to shy, awkward guys who lack the boldness to approach them. Second, these men exude self-confidence and status — key qualities that women desire in potential mates. They seek positions of leadership and social prominence, distinguishing themselves from the competition. Third, they dress well and stylishly, another status cue. Fourth, they have smooth and coordinated body language.[39] Their posture is relaxed and composed. They do not fidget nervously. They make good eye contact. More generally, they convey high mate value. They don't seem desperate. They don't trip over themselves trying to be nice or overly agreeable — qualities that can signal low status and submissiveness. They convey an assumption that they are fascinating and deserve to be at the center of the action. They attract women's gazes and boldly take social risks, commanding the group's attention — a key status indicator.

It's clear that men scoring high in Dark Triad traits have some of the keys to unlocking women's sexual psychology. But have individual men who score high in Dark Triad traits really won the battle? Or are women "victims" benefiting in some way? From an evolutionary perspective, women who succumb to the charms of high-scoring Dark Triad men could profit in principle in only a couple of ways. One might be what evolutionary biologists call "direct benefits" — reliable resources that aid in survival or reproduction. The status that these men seek and sometimes attain could confer tangible assets, such as food and shelter. The confidence these men exude could deter other men, yielding temporary protection to women and children from other men's attempts at sexual exploitation. A second class of potential benefits is genetic. Since there is no evidence that men high on the Dark Triad scale are healthier or have more robust immune systems, the women won't be getting genes for good health. They might, however, secure "sexy son genes," a hypothesis first advanced by the famous geneticist R. A. Fisher in 1930, long before the Dark Triad was discovered. If women bear sons who are high scoring in Dark Triad traits that make them more desirable to women, women can increase their reproductive success through the mating success of their sexy sons. Women won't have more offspring, but they could have more grandchildren fathered by sexy high-level Dark Triad sons who go on to use this alluring strategy to impregnate multiple women.

This is pure speculation, of course. We have no direct evidence for sexy son genetic benefits, although absence of evidence does not equate to evidence of absence. The relevant tests have not been conducted. The hypothesis would have to pass a high empirical hurdle, because these reproductive benefits would have to outweigh the costs that women vulnerable to high-level Dark Triad men are known to suffer, including reputational damage for being sexually exploited, physical damage from being abused, and loss due to being abandoned.

Research I conducted with Professor Peter Jonason discovered that men scoring high in Dark Triad traits are especially apt to use specific tactics for avoiding romantic entanglements with women after sex.[40] They keep their social worlds separate, intentionally failing to integrate the women into their networks of friends or family. They minimize intimacy, shun cuddling, avoid displays of emotional affection, and keep conversations light. Moreover, they flagrantly flirt with other women and boast about their past sexual successes — all actions that facilitate a short-term mating strategy by minimizing entangling involvements with any one woman. Their charming façade can fade quickly, revealing a nasty underlying personality. They are masters at ghosting.

High-scoring Dark Triad men tend to be disastrous as long-term mates, as we will see in Chapter 3. They lie, cheat, and steal. They poach other people's partners. They tend not to remain faithful. But they sometimes succeed in navigating the complicated maze of women's sexual psychology, at least for the short term. Young women are especially vulnerable to men scoring high in Dark Triad traits, perhaps because they are more likely to be seeking casual rather than committed mating, or perhaps because their inexperience leaves them more naïve and defenseless. High-scoring Dark Triad men make women feel vibrantly alive. They create exhilarating experiences that draw women into their bold and confident orbit. But short-term excitement often comes at a long-term cost, and being romantically rejected can have lasting effects, which we will explore later.

Men have no monopoly on being sexually dangerous, however. Women scoring high in Dark Triad traits also can be diabolical.

DANGEROUS DARK TRIAD WOMEN

Although men generally score higher on the Dark Triad scale, the sex difference is most pronounced on psychopathy. Men on average are more likely to agree with psychopathy statements such as "Payback needs to be quick and nasty"; "People who mess with me always regret it"; "I'll say anything to get what I want"; and "I like to pick on losers." Sex differences are minimal, in contrast, for the Dark Triad trait of narcissism (e.g., "I like to be the center of attention"; "I know that I'm special because everyone keeps telling me so"; "Those with talent and good looks should not hide them"; and "I am likely to show off when I get the chance"). Women are almost on a par with men on Machiavellianism, endorsing tactics such as: "Whatever it takes, you must get important people on your side"; "It's wise to keep track of information that you can use against people later"; "It's wise not to tell your secrets"; "Make sure your plan benefits you, not others."[41]

Women with high levels of Dark Triad traits are practiced in the art of sexual deception.[42] They use sex to obtain resources. They use sex to acquire clothes, money, good grades, companionship, or merely the gratification of sexual conquest. They parlay sexual favors to *maintain* access to partners who provide them with clothes and money. They are more likely than other women to fake sexual orgasm as part of a strategy to keep the resources flowing. Women high on the Dark Triad scale also are more likely than other women to say that they've had fewer past sexual partners than they really have had, perhaps because their actual numbers are typically higher.

Part of this strategy involves mate poaching — women high on the Dark Triad scale do a lot of it.[43] They feel no qualms about luring another woman's husband away for a sexual encounter just for the fun of the conquest. One woman described traveling by car through the German countryside and stopping into a village pub for a drink on her journey. She was the only female there. Several local men approached her, but she had already spied her target — a shy young man from the village sitting at the bar alone. He had recently gotten married. She approached him and after a few minutes of conversation suggested that they slip out back for a smoke. Outside she kissed him. His hands trembled, and she grasped them and assured him that everything would be fine. After a quickie, they agreed to meet later for more sex, away from the prying eyes of the village locals. But it was too late. Their encounter had not escaped the notice of his pub pals, who were worried about the damage it might cause his young marriage. They whisked him out the door and away from her. She got in her car and drove on with a satisfied smile. No guilt. No remorse. Just the quiet contentment of having conquered her prey.

Women high on the Dark Triad, we discovered, not only use mate poaching as a key mating strategy; they are particularly successful at it. They do not, like some, hold sex to be a misty-eyed romantic union of two souls. Sex is a powerful instrumental weapon, a means to get what they want, be it expensive clothes, gifts, money, or simply the satisfaction of capturing, however briefly, sexual prey. Like high-scoring Dark Triad men, they are skilled at the practice of deception, and whether it leaves a trail of broken hearts or broken marriages in their wake is not their concern.

With so much deception and sexual exploitation on the modern mating market, it's a marvel that people ever commit to long-term romance or agree to marry. Sexual conflict, as we will see in the next chapter, does not end with commitment or even with the wedding vows.

CHAPTER 3

STRUGGLES WITHIN MATESHIPS

> The male-female pair-bond [is afflicted] by the suspicions, jealousies, resentments, fears, anxieties, compromises, deceptions, disappointments, failures, nameless longings, named longings, misunderstandings, recriminations, divergent impulses, disparate fantasies, and conflicting moods — in short, the quiet desperation.
> — Donald Symons, *The Evolution of Human Sexuality*

HUMANS ARE NOTORIOUS AMONG ALL known species for their prolonged courtship — weeks, months, years. In sharp contrast, in many avian species, which share with humans a mating system of social monogamy and biparental care, courtship is brief — a few minutes of singing, a single dance in which the male must mimic the female's movements, or a twenty-minute inspection of the nest the male has laboriously constructed. Some people take years to make a commitment, courtship that requires a massive investment of time and resources. Courters incur opportunity costs, the forgone alternative potential partners who may be lost forever.

Given the enormous investment to arrive at commitment, it would be reasonable to expect that couple conflict should be minimal. After all, the newly mated couple has survived the grueling gauntlet of the mating market and made a careful partner choice. They've weeded out deceivers. They've had time to accurately assess each other's emotional stability, kindness, intelligence, ambition, career trajectories, and capacity for intimacy. They've had the opportunity to meet each other's networks of friends, which provide a wealth of social information. Many have secured evaluations from family members they trust. They've typically had time to test sexual compatibility, which can make or break the deal. They've evaluated the strength of the bond through stress tests, the slings and arrows of life, and intentionally imposed hurdles to see how much the partner is willing to sacrifice through thick and thin. And they've matched on mate value, more or less. In short, courtship seems exquisitely designed to produce long-lasting harmony. But is prolonged mating happiness an attainable goal?

THE EVOLUTIONARY RECIPE FOR MATING HARMONY

An evolutionary perspective provides a recipe for mating harmony and lifelong bliss, at least in theory. The first requirement is monogamy — long-term committed mating with zero chance of infidelity. Any chance of straying creates strategic interference, and with it, the evolutionary interests of men and women start to diverge. The second requirement is mutually produced offspring. Children become precious vehicles carrying the valuable genetic cargo of both parents into the next generation. Couples have a shared genetic stake in having their children survive and thrive. A third theoretical requirement is the absence of stepchildren. A mixture of jointly produced children and kids from a prior relationship creates the potential for conflict. What is in the best evolutionary interests of one parent departs from that of the other. A mother bringing a kid into the mix conflicts with the new stepfather, who has no genetic stake in the child; the same goes for a father bringing a child from a former mateship. The fourth requirement is zero chance of abandonment or breakup. Any possibility of leaving the relationship decreases the alignment of their interests, since reproductively relevant resources get taken away and diverted to other adaptive problems or to reproductive rivals.

The fifth requirement is an absence of prior entangling commitments, such as continued contact with an ex whose looming presence can siphon off time, energy, financial resources, and sometimes sexual resources. Even entangling social networks can create cumbersome costs — needy friends who bleed off time and emotional energy or hopeful backups who still carry a torch.

The final requirement for evolutionary harmony may seem the strangest — that each member of the couple must die at the same time. If one dies before the other, the living partner can re-mate. And re-mating opens the door to diverting acquired and inherited resources toward the dead mate's reproductive competitors. If the widow or widower re-mates within their reproductive years, additional children can be born that are entirely unrelated to the deceased partner. This, in turn, creates conflict and competition between the former mate's children and those born from the new mateship. This may be one reason that some women write inheritance wills that specify that resources go solely to their children, bypassing the husband entirely. In some cases women leave cash to the husband only on the condition that he not remarry.[1]

So we have the magic recipe for maximum strategic convergence of evolutionary interests and increasing the chances of mating harmony. But how often are all of these six key conditions met? Infidelities occur, sexual or emotional or both, violating the principle of monogamy. Some couples do not want to have children, and some that do cannot have them. Infertility hovers around 10 percent for reproductive-age couples who want kids and have unprotected sex for two years.[2] Couple infertility is a key cause of divorce in monogamous birds, and a key source of agony and strife in human couples.

Some partners come with kids from prior mateships, and this would have been true over human evolutionary history. Some fathers died in combat. Some mothers died in childbirth. And breakups often left a child with one or the other parent as the primary caretaker. Some potential mates come into the mating market burdened by needy friends or lingering hopefuls. Nearly all have entangling social networks, and it would be worrisome if someone lacked friends entirely. Given the 40–50 percent divorce rate in the United States, we know that commitment offers no guaranteed protection against defection.

And although committed couples are usually well-matched on mate value, discrepancies almost inevitably open up after commitment. An injury, illness, or accident — tragedies that evoke empathy for the afflicted — nonetheless can lower an individual's mate value temporarily or permanently. A promotion or demotion at work can create a mismatch where none previously existed. A family member from one side could suddenly require special assistance, moving in with the couple and saddling them with financial and emotional stress.

Women's mate value sometimes can rise dramatically. Dorothy Stratten's life trajectory illustrates this key point. She worked for minimum wage at a fast-food outlet in Vancouver. A small-time hustler named Paul Snider noticed her beauty, courted her, and became her boyfriend. He thought he could get her into *Playboy* magazine, so they moved to Los Angeles. It worked. She was featured in a series of photographs and spent time in the Playboy Mansion. She became a model and then an actress. This led to her meeting men considerably higher in desirability than Snider. Before long, she met Peter Bogdanovich, a famous young movie director best known for *The Last Picture Show*. Her acting career took off. She started an affair with Bogdanovich. She told Snider that she wanted to break up. Her star was rising. Although the story ended in tragedy, since she was brutally murdered by the spurned Snider, it proved to be a classic case of a woman trading up in the mating market when her status, fame, and mate value skyrocketed.[3]

An unusual act of heroism or bravery can also elevate mate value, as happened to a close colleague. Here is how he described the event:

> I was riding a long escalator down with [my wife] and then noticed that, on the adjacent escalator going up, there was a little girl on the outside of that escalator hanging on to the rubber handrail, going up. I looked up that escalator and saw that she would hit a glass wall at the very top. Somehow instincts took over and I raced down the escalator steps to get over to where she was heading. Sure enough, as she reached the very top of the escalator, she hit the glass wall and fell backwards. I barely got to where she was falling in time to kind of catch her. It wasn't a perfect catch as she ended up slipping through my arms, but she had a soft enough landing that she stood up immediately and started walking away. I received a lot of positive publicity from several newspapers and, in one paper, was one of 10 nominees for hero of the year.[4]

His social reputation and mate value shot up, increasing interest from women. His wife might well have ramped up her mate guarding.

Desirability, in short, never remains static. It drops with the slings and arrows of tragedy and rises with well-earned accomplishments and lucky blessings. It changes over the life course. Even fluctuations in a person's network of social alliances, such as the departure or death of a key ally or the cultivation of a powerful supporter, influence mating desirability and can create discrepancies within couples where none previously existed. Finally, what are the chances that both members of a couple die at the same time? It happens, but rarely. Re-mating is not at all uncommon after the death of a spouse.

Because something can always go wrong it would be surprising if people were caught entirely unawares and failed to develop anticipatory strategies. The first clue that people do anticipate, often unconsciously, comes in the form of cultivating backup mates.

MATE INSURANCE

When Professor Joshua Duntley and I first started to study backup mates, we did not know exactly what we would find.[5] We discovered something surprising. Not only do people in committed relationships have backup mates; even people who seem quite happy with their relationships actively cultivate them.

Consider this professional couple: Olivia and Noah,[6] a husband and wife, both successful, both bright, both attractive, both in their midthirties. They had met as medical students. Their love blossomed over common academic interests. They shared the trials and tribulations of progressing through a rigorous medical program, although they had different medical specialties. Olivia and Noah were so promising coming out of the gate that several prestigious hospitals offered them both positions. He concentrated on climbing the hospital's administration ladder, while she focused her energies on medical research. Once ensconced in their new positions, they started their family. Both delighted in posting Facebook photos of themselves with their two children in various states of play — frolicking in the snow, splashing in the rain, blowing out birthday candles. They also posted about each other's accomplishments, which were many, delighting in each other's professional successes. To all who knew them, they seemed the perfect and ideally matched couple.

Unknown to anyone, Olivia had grown unhappy. She slowly started to confide in a male professional colleague with whom she was collaborating on research. He lived in Norway. The colleague became a friend. One night while attending a medical conference abroad, after drinks and dinner, they fell into bed together. She became lovestruck. Her personal code of conduct, however, would not allow her to carry on a clandestine romance. What started out as collaboration and friendship blossomed into passionate love. She told Noah about her new romance and announced that she wanted a divorce. Noah was devastated. He had not seen it coming, had detected no discontent — perhaps because Olivia was a bit emotionally reserved, and perhaps because Noah was a bit psychologically clueless, with a mild touch of Asperger's. Noah plunged into a deep depression. He told Olivia that he still loved her and would wait for her, hoping she would have a change of heart, hoping her affair was transient. He wanted her back.

As several months passed, Olivia's long-distance love only grew stronger. She insisted on a divorce. With two kids, a former husband, and a new lover, she now had a five-body problem since she did not want her children to lose Noah's investment in them. In fact, she needed Noah not just for her children, but also so she could spend time with her new lover. In the meantime, after doing a bit of casual online dating, Noah unexpectedly found a new love — a woman who shared his passion for changing the hospital's culture to be more egalitarian. It was not a passing fancy, and the two even discussed having children together. When he revealed this to Olivia, she became enraged, barged into his apartment, and smashed some glassware. Noah was perplexed at her hypocrisy — after all, she had left him and was in the thrall of passionate new love, so why should she be so angry at him for finding happiness? Friends, too, were puzzled by her fury. Backup-mating psychology, it turned out, shed light on the puzzle. Noah had transitioned from being her husband to being her backup mate, a fallback should something go wrong with her new love. And Noah had said he would wait for her. Now that Noah had a serious new partner, Olivia had lost her most important backup.

Olivia's emotional reactions illustrate one of the interesting findings in our studies of backup mates. When we surveyed roughly three hundred people about how they would feel if their primary backup mate had sex with someone else, on a scale from "happy" (+3) to "upset" (−3), men more than women said that they would be upset, but only by a small margin. But when we asked how upset they would be if their backup fell in love with someone else, women were roughly twice as upset as men, with the average upset being −2.5. Similarly, women were more upset than men — extremely upset — if their backup entered into a long-term relationship with someone else. Olivia's loss of Noah as her backup explained what otherwise seemed like a hypocritical double standard — one standard for herself and a different one for Noah — and perplexing rage.

As it turned out, her loss of Noah as a backup came back to haunt her. Olivia's new lover secured a leave of absence from his job, which allowed him to leave Norway and join her. At last, their relationship could close the gap of distance. Long-distance romances often burn bright. Lacking exposure to the ups and downs of daily life, they allow people to fill in gaps in knowledge with positive illusions because people often overidealize a partner in the flush of a new romance. In the absence of direct information, they infer that the partner is honest, agreeable, loyal, and emotionally stable when they do not possess all the information needed for these inferences. Long-distance relationships sometimes allow people to avoid sleepless nights when the kids are up sick and crying and the sink is full of dirty dishes. The burps and body odors of real-life contact can take the sparkle off an idealized romance. The lover who burns so brightly in a fancy hotel in a faraway glittering city must surely be the kindest, most intelligent, most thoughtful, most sexually erotic, most fascinating person on earth. Within a month of the couple finally being united in one city, Olivia's lover jilted her unexpectedly. He didn't want to be saddled with the burden of her two kids. He didn't want to disrupt his European lifestyle. And he had found a new flame closer to his home. It was Olivia's turn to become depressed. Without Noah as her backup, with two dependent children, with her divorce final, she now had nowhere to go and no one to turn to.

Women and men both cultivate backup mates, and they serve several key functions. First, they can be possible replacement mates should something catastrophic happen to a regular mate. Second, backups can serve temporary functions, providing psychological support or sexual fulfillment during the transition back into the mating market. Breakups inflict blows to self-esteem; backups can offer a boost. Third, backups offer protection and provisions, shields against the hostile forces of nature. It is for these reasons that the loss of a primary backup is so upsetting. The mating safety net gets yanked out. The comforting bridge back to the mating market collapses. Backups render people less vulnerable, and the loss of a backup can be calamitous. We found that people without a backup mate were twice as likely to get depressed, as gauged by the Beck Depression Inventory, compared with those with a solid backup.

Backup mates, in short, provide a bounty of benefits. People actively cultivate them, sometimes consciously, but often subconsciously. They usually conceal them from their regular mates. These backup mates create couple conflict if discovered. Both sexes experience depression and psychological angst if a backup forms a serious romantic attachment with someone else. The idea that even happily committed couples form backups "just in case" may put a large dent in our fantasies of the purity of love and romantic visions of lifelong romantic harmony. But backups were essential in providing mate insurance to our successful ancestors. They remain necessary in the modern world of mating, where the best-laid plans can spiral off course unexpectedly.

WHY DO MEN AND WOMEN HAVE AFFAIRS?

Among married couples, infidelity is far from a trivial occurrence.[7] In 1953, Kinsey estimated that half of all married men and a quarter of married women experienced at least one affair, although other studies put rates lower or higher. We *know* that infidelity is the leading cause of divorce worldwide, from the Inuit in Alaska to the !Kung San of Botswana. And we *know* that most adults in the modern world, including roughly 85 percent in the United States, have experienced at least one romantic breakup, and infidelity is one key cause.

Why most men have affairs, from an evolutionary perspective, is fairly straightforward. Men have evolved a strong desire for sexual variety, stronger than women's on average, due to the large asymmetries in parental investment. Men can reproduce with as little effort as it takes to inseminate a fertile woman, although men typically invest more than the minimum. Women require a metabolically costly nine-month pregnancy to produce a single child. Breastfeeding an infant, which lasts roughly two to four years in traditional cultures, compounds the costs women incur. Stated differently, an ancestral married man with two children could have increased his reproductive output by 50 percent by a single successful reproduction with an affair partner. Adding sex partners for women who already have one generally does not, and never could have, dramatically increase their reproductive success. Men do not think about the reproductive logic of their actions, of course. They just find other women to be sexually attractive, sometimes get sexually bored with their current partner, and may succumb to sexual opportunities if the costs and risks are low. Although their motives for infidelity do not always boil down to the sheer joys of a novel sex partner, the desire for sexual variety is overwhelmingly what propels most men to stray.

But there has always been one missing piece of the puzzle

when it comes to understanding women's mating strategies — a gap in scientific understanding of why women have affairs. Until recently, evolutionists have typically explained female affairs with the *good genes hypothesis:* the idea that women prefer affair partners with traits associated with increased offspring survival, such as a super-healthy immune system, and therefore "good genes." The hypothesis posits that women have evolved a dual-mating strategy — securing investment from one man while mating on the side with men who have better genes than their regular partners. Although many evolutionists endorse the dual-mating strategy as an explanation for why women have affairs, I believe there is a more compelling explanation.

The *mate-switching hypothesis,* I believe, provides a scientifically supported answer — that women have affairs to extricate themselves from a poor mateship and trade up to a better partner. For both sexes, the hypothesis explains what we commonly observe: Sometime after publicly declaring her marriage vows, a woman finds herself sexually attracted to her co-worker. After changing his child's fifth diaper of the day, a man wonders whether he made a terrible mistake and fantasizes about the high school sweetheart who got away. After years of marriage, a woman finds that she's the primary breadwinner, and her husband's laziness has eroded her confidence in their union; she notices that her co-worker lingers longer in the doorway of her office than usual. After years of living a life of unfulfilled desires, a man starts a passionate affair with his next-door neighbor. A woman confesses to her best friend that she's in love with another man and secretly lays the groundwork for leaving her husband — a separate bank account and a deposit on an apartment.

These diverse examples stem from a common cause — humans have evolved strategic adaptations for mate switching, a phenomenon that is widespread across species. The simplest such adaptation is deployment of a "walk-away" strategy, in which people physically separate themselves from costly cooperative partners. The mate-switching hypothesis proposes a version of the walk-away strategy underpinned by psychological adaptations designed to detect and abandon costly mates in favor of switching to more beneficial mates.

Many in modern cultures grow up believing a myth about life-long love. We are taught about falling for the "one and only." We learn that the path to fulfillment is paved with a single glorious union. But the plots of fictional love stories typically end upon the discovery of that one and only; they rarely reveal the aftermath. The story of Cinderella ends with her getting the prince. After overcoming countless obstacles, a union is finally consummated. Few romantic fantasies follow the story line of committed mating — the gradual inattentiveness to each other's needs, the steady decline in sexual satisfaction, the exciting lure of infidelity, the wonder about whether the dull grayness of marriage is really all life has to offer.

In fact, we come from a long and unbroken line of ancestors who went through mating crises — ancestors who monitored mate value, tracked satisfaction with their current unions, cultivated backups, appraised alternatives, and switched mates when conditions proved propitious. To understand why, we must turn our gaze to those ancestors and uncover the mating challenges that they confronted.

Ancestral humans faced three great struggles of life. First were the hazards of the physical environment — getting enough food to eat, finding shelter from storms, fending off extremes of heat and cold. Second were struggles with other species. Survival was always at risk from dangerous snakes, carnivorous cats, and parasites that made our bodies their homes. A third class of challenges proved no less fundamental — competition and conflict with members of our own species. Other people, with the multifarious tactics they use to trip up their reproductive rivals, collectively made up a momentous hostile force of nature.

In the context of these recurrent struggles, humans evolved a menu of mating strategies, of which long-term committed pair-bonding became central. A committed mate could provide meat during cold winters when no berries were blooming. A long-term partner could offer protection from hungry predators and hostile humans. Life mates could nurture one's children, the invaluable vehicles that carried their genes into the future. Long-term mating, in short, offered a bounty of benefits, aiding in combat against all three classes of human struggles.

But as we have seen, something could always go awry. An initially promising hunter could get hobbled by injury or infection. A regular partner could get bitten by a poisonous spider, wounded in battle, or killed in intergroup warfare. Or a mate's status within the group could plummet, decreasing privileged priority for access to the group's critical resources. A partner's mate value, in short, initially promising an upward path, could suffer calamitous setbacks. Long-term mate selection is all about future trajectory, and the future often carries with it treachery and tragedy.

Another challenge facing a committed mateship is that more valuable mates, initially not present or unavailable, sometimes appear on the scene. A previously unavailable potential mate could suddenly become unencumbered due to the death or desertion of their own partner. The merging of two separate tribes could present a fresh profusion of mating opportunities. In short, the vagaries of life provided new prospects for our ancestors to trade up in the mating market.

All of these ancestral challenges favored the evolution of strategic solutions. Some solutions involve tactics of mate retention, motivations to fend off mate poachers and hold on to an investing partner. These tactics range from vigilance to violence. But there existed another important suite of solutions — adaptations for mate switching.

Although much scientific research has focused on the initial stages of mate selection and mate attraction, and some on mate retention, relatively little attention has been given to adaptations for mate switching. One of the most important involves monitoring a partner's mate value, which is made up of dozens of qualities, as discussed earlier. These include social qualities such as the status or esteem in which they are held, their networks of allies, the power of their close kin, and many others.

Personality is important too. Is a partner energetic, dependable, ambitious, emotionally stable, sociable, easygoing, or dominant? Most of these qualities change over time. Social status can rise or fall. Health and well-being rise and fall day-to-day but can also be impaired more permanently by a parasite, disease, or injury. Personalities change. Energy levels can ebb with age. Ambition might wane after mate selection. Even emotional stability can change due to psychological or physical trauma. Post-traumatic stress disorder (PTSD) is a common consequence of the ordeals of war and sexual assault — a topic discussed in detail in Chapter 8. Changes in these key components of mate value, inevitable in all but the most insulated lives, must be monitored.

A partner's mate value is critically also a function of how much they value you. The technical term is welfare trade-off ratio (WTR), the ratio of how much value they place on your welfare relative to their own welfare. Some mate selectors suffer a rude shock when an altruistic WTR during the courtship phase turns into a selfishly skewed WTR after the wedding vows. This might be one reason why divorce is most common in the first few years of marriage and then tapers off over time. A partner who initially shows high investment might curtail that investment after months or years — a form of resource infidelity discussed later in this chapter. Relationship satisfaction, a barometer that goes up and down with the tides of time, is the key psychological monitoring mechanism that tracks components of a partner's mate value, their level of investment, and the WTR they hold with respect to you.

Mate value within couples is inherently relative. Consequently, monitoring a partner's mate value is not enough. Self-assessment is required. A woman's or man's mate value can increase over time. Either person might rise in status, inherit a bounty of resources, or distinguish themselves through acts of bravery, leadership, or wisdom, rendering them more desirable on the mating market. A woman whose mate value increases can find herself dissatisfied with her husband, even if his overall desirability has remained unchanged, as we saw in the tragic case of Dorothy Stratten and Paul Snider.

Women high in desirability elevate their mating standards, expecting higher levels of mate qualities on metrics such as status, resources, commitment, and cooperativeness. A woman's mate value even varies over the monthly ovulation cycle. Subtle changes in women's attractiveness reflect the ovulatory phase. Their skin glows a bit more, their waist-to-hip ratio becomes slightly lower, and their voices rise a bit, all qualities found to enhance perceptions of female beauty. The fact that women become more exacting in their mate preferences at precisely this time in their cycle might reflect an adaptation to monitor their own mate value and adjust their standards accordingly.

We do not know whether these cyclical or more enduring changes in women's desirability influence qualities such as their level of satisfaction with their current partner, their attraction to alternative potential mates, their effort to cultivate backup mates, or their temptation to have an affair. But there is tantalizingly suggestive evidence. One study found that women are most likely to evade their partner's mate-guarding efforts precisely around their most fertile phase — an effect most pronounced among women partnered with men low in attractiveness.[8] These women find themselves more interested in attending social events, perhaps because they might interact with alternative mates. And they report engaging in greater flirtation with men other than their regular partners. The findings point to the possibility that women monitor their own mate value, and when it increases, alternative mates can seem more attractive. This, in turn, requires monitoring alternative possible mates.

According to the mate-switching hypothesis, scanning for alternative mates remains activated even among those in happy relationships. Sometimes that tracking occurs at low levels when newly available mates appear or when a potential mate's attraction or interest increases. Sometimes it gets activated at high levels, as when a woman becomes increasingly dissatisfied with her regular mate and wants out of the relationship.

According to our studies, led by Professor Daniel Conroy-Beam at the University of California, Santa Barbara, and Professor Cari Goetz at California State University, San Bernardino, women tracking alternative mates operate in three dimensions: The first is *interest*; does the potential mate show attention, attraction, and desire? Prolonged eye contact, selective smiles, and sideways glances are some documented indicators here. Do these indicators signal long-term interest or a fleeting sexual desire? Most women will not leave their regular partner for a passing fancy, although some see it as an important signal that something is seriously wrong in their regular relationship. The second dimension is *mate value*; only large increments in value over the value of the current partner are likely to be worth the costs of breaking up. The third dimension for tracking is *eligibility*. Is the interested alternative actually free of encumbering commitments such as an existing spouse or the crushing commitments of dependent children? Conroy-Beam, Goetz, and I found that women scaled back on the effort to retain their regular partner only when operating in one or more of the dimensions above.

These widespread patterns would never have evolved without producing real-world mating decisions and behavior. So how do women actually implement a potential mate-switching strategy? My colleagues and I propose three key strategies — cultivating backup mates, a key concept we've already introduced; implementing affairs; and enacting a breakup.

Infidelity is an effective tactic for prompting a divorce, but it is also dangerous. In fact, infidelity is the leading cause of intimate partner violence and is also a key motive behind spousal murder — topics we explore in depth in Chapter 5. Despite these risks, many women, roughly 20–30 percent, take the plunge and launch into an affair. Interestingly, married women in their early thirties are most likely to have an affair, perhaps reflecting a motivation to switch mates while their desirability is high and they are still fertile.

Additional lines of evidence support the notion that infidelity serves a mate-switching function. First, women who initiate affairs are much more likely to suffer from marital dissatisfaction than women who do not. This might seem blindingly obvious, but the same studies show that men who have affairs do not, in fact, differ from those who abstain from affairs in their levels of marital happiness. Second, women are much more likely than men to become emotionally involved with, and to fall in love with, their affair partners. Roughly 79 percent of women report doing so, in contrast to only 30 percent of men.[9] Moreover, women are more likely to cite emotional involvement as a reason for the affair. Men are more likely to cite pure sexual pleasure. These critical sex differences point to dramatically different functional reasons for male and female infidelity. For women especially, they point to the mate-switching function; for men, the desire for sexual variety.

Exiting a current mateship is the final step. One clichéd ejection tactic involves saying "It's not you, it's me," in an attempt to minimize rage-motivated vengeance. A second involves transforming the existing romantic relationship into a friendship, also designed to minimize the ire of the ex. In some cases, women might continue for a time to provide sexual favors to an ex to minimize the harms he might inflict or keep him as a backup mate. Alternatively, a woman might attempt to direct his sexual attention to other women. The effectiveness of these and other ejection tactics, and the circumstances in which each tactic is deployed, remain to be studied scientifically.

The mate-switching hypothesis explains an array of findings that otherwise remain mysterious. It explains why people cultivate backup mates and why desired qualities in opposite-sex friends closely mirror the desired qualities in a long-term mate. It explains why women become dissatisfied with their existing relationships when there are available alternatives in the mating pool who surpass their regular partners in mate value. It provides a cogent explanation for why women are willing to risk so much to have affairs, an enduring evolutionary puzzle because women gain no direct benefits in the currency of reproductive success. It comes as no surprise that some women are especially likely to take these risks — women who score high on the Dark Triad traits of narcissism, Machiavellianism, and psychopathy.[10]

Recognizing that humans have evolved a psychology dedicated to mate switching undoubtedly will be disturbing to many. It might be disconcerting to a man to realize that his wife is carrying a mate-insurance policy, harbors sexual fantasies about her co-worker, or has "just a friend" who is more his rival than he realizes. It might be depressing to know that you are more replaceable than you knew. It could be disturbing when it dawns on you that your partner's unhappiness might not be transient and instead portends a hidden plan for exiting the relationship.

But nothing in mating remains static. Evolution did not design humans for a half century of matrimonial bliss. Few of our ancestors lived past the age of forty. Those who stuck it out through thick and thin might win admiration for their loyalty. But modern humans have descended from successful ancestors who carried mate insurance; who devoted energy to scenario building, the cognitive simulations of fantasizing about possible mates and laying plans for exiting; and who acted on those scenarios when the hidden calculus pointed to the benefits of switching mates.

Sexual infidelity, cultivating backup mates, and the subterranean psychology of mate switching are not the only causes of sexual conflict within mateships. Another revolves around the diversion of pooled resources.

RESOURCE INFIDELITY AND SEXUAL CONFLICT

The Ache of Paraguay are foragers who survive by hunting and gathering. Hunting skills are paramount for a man's status within the group and his desirability to women.[11] Poor hunters struggle to attract a wife. Skilled hunters attract two, three, or more. The wife of a talented hunter, however, gets no more meat than anyone else in the group. Ache social rules dictate that a successful hunter must deposit his kill with a central distributor — typically a wisewoman known for fairness. Everyone in the Ache shares the meat equally. Or so it seems on the surface.

In fact, talented Ache hunters typically have a mistress or two. Before returning to home base with the kill, the skilled hunter slices off a prize tender cut of meat. Then he dispatches an emissary to deliver the meat to his mistress before returning home with the remaining bounty. This is an ancient form of resource infidelity — the diversion of pooled resources to an extra-pair partner. In modern cash economies, it takes the form of financial infidelity.

In today's Western cultures, mated couples often combine resources. Couples use jointly held resources for mutually beneficial expenditures — a house in which they live, food for the family, a child's education, a TV that all can watch, a house wired for internet that everyone can access. However, when the evolutionary interests of the woman and man diverge, such as when it is advantageous for one partner to have an affair or trade up in the mating market, the possibility for resource infidelity opens up.

How common is resource infidelity? One survey of a thousand individuals in New York City revealed that 40 percent of married women and 37 percent of married men had a secret bank account.[12] A Harris survey found somewhat lower numbers, but nonetheless 31 percent admitted lying to their spouse about some aspect of finances — 58 percent of those who did hid cash, 30 percent hid a bill, 15 percent hid a bank account, 11 percent lied about earnings, and another 11 percent lied about debt.[13] Another study found that 80 percent of respondents admitted to hiding money from their spouse.[14] Regardless of the exact numbers, which vary from study to study, sample to sample, and method to method, financial infidelity is clearly not uncommon.

Every comprehensive study of sources of struggles in couples reveals that financial conflicts are prominent. They are rivaled in frequency only by conflicts about sex. Financial disagreements within committed couples are among the most persistent, recurrent, and heated. Importantly, they are the least likely to be calmly resolved.[15] Quarrels over money tend to evoke anger, hostility, and tactics of conflict resolution that are destructive psychologically and sometimes physically. Financial disagreements turn out to be one of the strongest predictors of divorce.[16] Despite wedding vows that invoke remaining together "for richer and for poorer," money conflicts lead to breakups. Sometimes these conflicts center on insufficient provisioning or economic hardship. When the size of the pie is small, it is understandable that nerves fray. Often, however, the conflict centers on how the pie is divided, who controls the size of the slices, and who conceals the size of their slice.

An evolutionary psychological analysis yields insight into why financial struggles loom so large in marital conflicts. A key clue comes from gender differences in spending. Men are more likely to splash out cash for status-enhancing entertainment systems, new technologies, and flashy cars.[17] Women are more likely to spend on designer clothing and appearance-enhancing cosmetics. In a nutshell, each uses pooled resources to enhance sex-linked components of their mate value — conspicuous displays of status for men and beauty-boosting products for women.

Men's conspicuous consumption derives from an evolutionary insight known as the *handicap principle*. The logic behind the handicap principle, originally proposed by the Israeli biologist Amotz Zahavi, is that only those who have abundant resources can afford to squander them.[18] Spending on luxury goods, therefore, can be interpreted as an honest cue — a signal that one possesses a bounty that others lack. It is an honest signal because those who lack resources cannot afford to waste money on needlessly extravagant items. A Tesla that costs $100,000 or a Rolls-Royce that costs $350,000 does not get a man from point A to point B any faster or more efficiently than a humble Honda that costs $24,000. But assuming the money is not stolen or borrowed, luxury cars signal a level of resource abundance via the handicap of wasteful spending that cannot be faked by those of poorer means. It increases a man's perceived status and economic position — universal components of his mate value.

Women's conspicuous consumption serves different functions. The first is appearance enhancement. Women vastly outspend men on cosmetics — toners, primers, eye creams, moisturizers, concealers, bronzers, eye shadows, mascaras, blushers, eyeliners, and lipsticks. Makeup works. It increases perceptions of beauty.[19] Women are also more likely than men to spend on cosmetic surgery — breast implants, tummy tucks, and Botox injections. Women's expenditures, in short, function to increase perceptions of their physical beauty, which is a more important component of female than male mate value.

The use of pooled resources to enhance one member of a couple's mate value can lead to a raft of problems. It can create mate-value discrepancies where none previously existed. A woman's newly enlarged breasts, bee-stung lips, and tummy-tucked waist render her more desirable to a larger pool of men. A man's extravagant set of wheels makes him more desirable to a larger array of women. Mate-value discrepancies produced in these ways can lead to an arms race in which each partner struggles to siphon off mutually held resources to keep their desirability on pace with the other's. If the end result is equality in mate value, they end up right where they started. Both would have been better off if neither wasted the money, and instead used it to deal with challenges they share together, such as investing in their children.

Mate-value discrepancies, however, are far from the only problem caused by the financial struggle. Elevated status, even in the absence of a desirability discrepancy, makes men more attractive to women as casual sex partners, creating more numerous temptations to be unfaithful. Creating sexual opportunities may be precisely the point for men. One study found that men who are inclined toward short-term mating are most likely to purchase luxury cars, high-end technology, and other status-enhancing displays.[20]

Long-term-oriented men, in contrast, are more likely to opt for the practical, reliable car lacking extravagant flash. Facebook founder and billionaire Mark Zuckerberg drives a $30,000 Acura — safe and not ostentatious. Billionaire Warren Buffett drives an older model Cadillac that cost only $24,500. John Mackey, multimillionaire founder and CEO of Whole Foods, drives a Honda Civic. As far as we know, all three are happily married long-term maters.

Dark Triad traits also enter the equation. A seminal study rated photos of women and men with makeup removed, dressed only in a plain gray T-shirt, with hair pulled back if they had long hair, and providing a neutral facial expression — *unadorned attractiveness*.[21] Photos were taken of the same people as they showed up for the laboratory experiment in whatever clothes and makeup they happened to be wearing — *adorned attractiveness*. A panel of unacquainted judges rated the attractiveness of each photo. Dark Triad traits were not correlated with unadorned attractiveness, meaning that those who score high on narcissism, Machiavellianism, and psychopathy are not intrinsically more attractive than those who score low on those factors. High-scoring Dark Triad individuals, however, are rated as significantly more attractive in the adorned condition; that is, they use more makeup, more expensive clothing, and more bling to elevate their attractiveness. Conspicuous consumption, by increasing attractiveness, enables those high in Dark Triad traits to more successfully implement a short-term sexual strategy.

Women accurately infer that men who conspicuously consume are likely to be interested in short-term mating. This inference, combined with greater temptation from interested women, creates a profound adaptive problem for wives—the problem of mate retention.

Women's mate-retention tactics are multiple. The first involves keeping the partner sexually satisfied. Among the Muria tribe of Chhattisgarh, India, for example, wives often insist on sexual intercourse before the husband leaves the house in the hopes that it will reduce his temptation to succumb to other women.[22] A woman's conspicuous consumption may serve a similar function by rendering her more desirable to her mate. The more attractive she is, the more sexually interested he will remain, and the more effort he will expend on keeping her committed to him. A study of fifteen hundred married individuals found that men who perceive that their wives are quite attractive, in contrast to men who do not, maintain a high frequency of sex throughout the marriage and report high levels of sexual satisfaction.[23]

Conspicuous consumption also sends a signal to other women that her partner is heavily invested in her.[24] It improves her desirability in comparison with other women. Female mate poachers look for men who might be vulnerable to being lured away. Women's conspicuous consumption, focused on appearance enhancement, signals to their rivals that their mates are securely "taken" and not easily poachable.

SEXUAL DOUBLE STANDARDS AS SOURCES OF CONFLICT WITHIN MATESHIPS

We saw in Chapter 1 the persistence of a sexual double standard on the mating market, even in sexually egalitarian countries like Norway. But does a double standard persist once a mateship has already formed? The stereotypical double standard holds that it is acceptable, or at least not that morally shameful, for husbands to have sex outside of marriage, but it is not acceptable for women to do so. Women who have experienced this double standard express genuine puzzlement. It seems logically inconsistent, they reasonably argue, for a man to think it is okay for him to cheat but not so for his wife. This double standard is indeed *logically* inconsistent at a level of abstraction that holds what's good for the goose should be good for the gander.

But in men's conscious and unconscious minds, a sexual double standard is not at all *psychologically* inconsistent. The reason is that men's double standard stems from at least two distinct sexual adaptations. The first is the desire for sexual variety, a strong attraction to novel women. This desire, as we have seen, is one of the strongest and most replicable findings in sex research, and indeed in the entire field of psychology. Men use their desires to justify or rationalize their affairs. As Kevin Kline says in character, playing a husband who frequently cheats on his wife (played by Tracey Ullman) in the movie *I Love You to Death*, "I'm a man. I got a lot of hormones in my body."[25] A real-life rationalization of this desire was eloquently expressed by Joseph Smith, founder of the Mormon religion: "Monogamy seemed to him — as it has seemed to many men who have not ceased to love their wives, but who have grown weary of connubial exclusiveness — an intolerably circumscribed way of life. 'Whenever I see a pretty woman, I have to pray for grace.' But Joseph was no careless libertine who could be content with clandestine mistresses.... He could not rest until he had redefined the nature of sin and erected a stupendous theological edifice to support his theories on [plural] marriage."[26]

The Lord tells Joseph: "And if he have ten virgins given unto him by this law, he cannot commit adultery, for they belong to him.... But if one or either of the ten virgins, after she is espoused, shall be with another man, she has committed adultery, and shall be destroyed... according to my commandment."[27]

Despite men's rationalizations for imposing a sexual double standard about infidelity, Joseph Smith's sexual psychology reveals precisely why it does not seem to him to be at all inconsistent. He shares with most other men a powerful desire for sexual variety, and with men high in power and social status a strong sense of sexual entitlement to act on those desires. And also he shares with most other men adaptations, the cardinal one being sexual jealousy, that function to control his wife's conduct and to monopolize all of her sexual assets. These dual sexual adaptations explain the mystery of why so many men find sexual double standards about their conduct and their wives' to be perfectly psychologically consistent, regardless of whether the wives also experience a desire for sexual variety. It explains why "her rational brain wonders why if he's so relaxed about his wandering appetites, he should become suddenly so anxious if her appetites start to wander too."[28]

A deeper look reveals that gender differences in sex with other partners comprise only one sexual double standard. An intriguing study posed this question to men and women: "What counts as sex?"[29] Only 41 percent of the men in existing relationships said that oral contact with someone else's genitals would count as sex. But 65 percent of the men said that if their partner had oral contact, it would count as sex. Does this reveal the usual sexual double standard, where women are evaluated more harshly than men for the same conduct? The answer resides with what women say. About a third of women, 36 percent, indicate that if they had oral contact with someone else, it would count as sex — about the same as what men say. But 62 percent say that if their partner had oral contact with someone else, it would count as sex. These findings reveal a previously unexplored sexual double standard — not between men and women as groups, but rather between standards people hold for themselves versus their partners — what we can call the *me versus thee* double standard.

If people hold sexual double standards about what counts as sex — not sex if I have contact with others, but definitely sex if you do — it is easy to see how this psychological quirk can lead to sexual conflict within relationships. It's okay for me to kiss someone else; it doesn't really mean anything, and besides, it's not really sex. But you had better not. It's okay for me to receive a bit of oral pleasure when you're out of town because it's not really sex. But if you do, it's infidelity with a capital *I*.

In addition to sexual double standards for oneself versus one's regular partner, women and men actually define infidelity differently. For men, infidelity is almost entirely about sex. They define infidelity quite narrowly. Upon discovery of a mate's involvement with another man, the most common verbal interrogation is "Did you have sex with him?"[30] Here is one woman's account of how her boyfriend reacted: "My boyfriend and I broke up, but then we decided to rekindle the relationship. He insisted on knowing what I'd been doing since our first break-up, and I told him how many people I'd slept with. He was *very, very, very* upset. He had come to visit for the 4th of July, and I thought we would rekindle what we had. Boy did I ever blow the spark out of that one! When I told him the number of people I'd slept with he nearly jumped through his skin."

Women endorse a broader definition of infidelity. They include sexual intercourse with someone else, of course. But women define it as also encompassing *emotional infidelity.* Here is one woman's account: "My boyfriend got a call from a girl that used to like him not that long ago and they were talking up a blast. He was laughing and sounded really interested in what she was saying and cracking jokes with her...etc. The fact that we had not had a conversation like that in a really long time and we were supposed to hang out that night, and while he was on the phone, I just sat there waiting for him...I yelled at him when he got off and refused to have sex that night."

The betrayal is not obviously sexual, but it involves psychological intimacy with another woman. It involves her boyfriend diverting the precious resource of attention. And the act of sharing humor and emotionally intimate moments with another woman warrants withdrawing sex — a clue that we must follow to understand another key form of sexual conflict within relationships.

SEXUAL WITHDRAWAL AND BESTOWAL AS LEVERS OF POWER

"Everywhere sex is understood to be something that females have that males want," according to evolutionary anthropologist Donald Symons. He is only partly correct, and we examine that part first. But as we will see, Symons ignores the fact that sex is also something that women in relationships desire and that they become upset when it is withdrawn, albeit for somewhat different adaptive reasons.

Once people have pair-bonded after a prolonged courtship, it is sometimes taken for granted that each partner has more or less continuous sexual access to the other partner. In many couples and in most cultures, it is expected. In ancient Israel, for example, marriage without the expectation of sex was incomprehensible.[31]

Every married couple was expected to fulfill sexual obligations and to reproduce. Marital alliances typically institutionalize exclusive sexual and reproductive entitlements, and this is true in all known and well-studied cultures, including traditional hunter-gatherers.[32]

The sexual entitlements men gain from marriage are reflected in laws regarding adultery. Historically, adultery by the wife was considered a "property violation," with the male mate poacher illegally encroaching on the exclusive sexual access granted to the husband through the marriage contract. In seventeenth-century English law, for example, adultery was seen as a capital crime in which both parties warranted punishment.[33] It was analogized to a property crime in which sexual infidelity constituted theft. The aggrieved husband was due financial recompense, and the wife's infidelity was legal grounds for divorce. These old laws were written by men, of course, and so reflect a key component of men's sexual psychology. Most modern Western cultures have eliminated these laws, although they continue to exist in most Islamic cultures. Western laws now give married women control over their own bodies, and many cultures have instituted laws against marital rape — a topic we take up in Chapters 7 and 8. But the male sexual psychology that gave rise to the laws to begin with — specifically male sexual proprietariness — continues to be fully activated within committed relationships.

Cultural shifts toward greater gender equality within relationships have dramatically reduced men's entitlements. Western marriage no longer grants men unconstrained sexual access whenever and wherever they want. Women within committed relationships have the rights and freedoms to consent to sex or to withdraw sex. And this gives women a critical lever of power — the power to reward and the power to punish.

One female advice columnist made this explicit: "You can use sex as a bargaining tool to make your man act right in the relationship. Men are simpleminded beings that can be controlled like puppets. A man will do virtually anything on the promise of sex. I've seen men drive for miles, skip a job interview or an exam and put up with psychotic women just to get laid."[34] The advice goes on to recommend using sexual reward and withdrawal to get the man to do the dishes, take her out on a romantic dinner date, and even to buy her a new car.

When world-renowned sex researcher Professor Cindy Meston and I explored the many reasons that women have sex, we found that some women in relationships use sex as pellets of reward. One woman said, "I have sex to get my way or to persuade my husband into something I really want that he might be opposed to." Sex can be used for rewards big or small. One woman said, "My boyfriend bought me a car a few years back. I wasn't in the mood for sex, but he was, so we had sex whenever he asked...for at least a few weeks." Another said she had sex "so that he'd take out the damned garbage."[35]

Others think this is terrible advice. Dissenters argue that using sex to reward and punish will create more fights, cause men to withdraw emotionally, and even drive them into the arms of other women or give them an excuse to cheat. As one woman put it, "Whatever you don't do, another bitch will be more than willing to do it."[36] Clearly, using sexual withdrawal as punishment or sexual favors as reward must be done skillfully. It's not a universally effective tactic. If the man has other sexual options or if he's higher in mate value, for example, the tactic could easily backfire.

But it's clear that for some women in some relationships, it works. One reason goes back to sex differences in sex drive. Although there is much overlap in what men and women want when it comes to sex, evolution has equipped men with a higher sex drive. This is reflected in several sexual adaptations. Men become sexually aroused more easily than women, especially to visual stimuli such as a woman wearing skintight or revealing clothing. They have more frequent spontaneous sexual fantasies than women. They spontaneously think about sex twice as often as women do every day.[37] They desire to have sex more frequently than do women. Testosterone, a key hormone linked to sex drive in both genders, shows an enormous gender difference. Men average 679 circulating units of testosterone; women's average is a tenth of that.[38] In short, gender differences in sex drive create a gap. The less interested person often has more power over if and when sex will occur, and women are often less interested.

But not always, and this is where the opening quote by Donald Symons misses a critical sexual dynamic. A study of 966 Zimbabwean women found that 17 percent of married men actually withdrew sex from their wives, and women were extremely unhappy about it.[39] Women's upset turned on the predictable circumstances in which it occurred. The first centers on concern that the relationship might end. Women worry that the man's withdrawal of sex, especially when they are eager to have sex, signals the loss of love. It portends the loss of his investment, putting their children at risk. And it is a foreboding sign of separation or divorce. Of the women in the study who were separated or divorced, 27 percent reported that their partner stopped having sex prior to the separation.

The other key cause of men's sexual withdrawal stems from the husband taking on a girlfriend on the side or seeking to add a second wife. One woman put it this way: "I think another form of abuse obviously is when the husband has a girlfriend, he will make it so obvious to the wife that he does not really care for her because he has somebody better, but he would want to keep this wife as if she is just a house maid, because the wife will do the cooking, the washing of his clothes. He has somebody whom he takes out and has a good time with, but the wife is deprived of what should be part of the marital set-up, the relationship that they are supposed to have as man and wife."[40] The man's sexual withdrawal portends the loss of love and its redirection to another woman. Or, if the wife refuses to allow her husband to take on a new lover, he might punish her by refusing sex with her.

Both women and men want partners who appreciate their sexual relationship and who find them sexually alluring. Withdrawal of sex can signify that your partner does not find you sexually desirable. It's sometimes a sign of a decrement in your partner's perception of your mate value, which is why both women and men find their partner's sexual withdrawal so troubling.

Despite the on-average sex difference in sex drive, the allocation or withdrawal of sexual resources can be used by both sexes. Sex can be dispensed as a reward when a partner does something you desire, and withdrawn to punish a partner for doing something you dislike. And a key part of its worth in sexual skirmishes comes from the ability to toggle back and forth, to bestow or withdraw, which maximizes its perceived value. It is precisely this variability that functions to render sex a scarce resource.

This chapter has covered some of the major struggles within mateships — struggles that center on sexual infidelity, financial infidelity, sexual double standards, and the use of sex as a lever of power. We have seen that sex can be used as a weapon to reward or punish a partner, and that sexual withdrawal sometimes portends a breakup. The next issue is how couples cope with sexual conflict — a topic to which we now turn.

CHAPTER 4

COPING WITH RELATIONSHIP CONFLICT

> Sometimes, God doesn't send you into a battle to win it; he sends you to end it.
>
> — Shannon L. Alder

THE ADAPTIVE CHALLENGES POSED BY relationship conflict have recurred generation after generation over human evolutionary history. Our pair-bonded ancestors commonly encountered mate-value discrepancies, resource infidelities, threats from mate poachers, and sexual infidelities. The costs of being on the losing end of these challenges created selection pressures for effective solutions. Humans have evolved tactics for minimizing the odds of encountering these problems and for solving them in a successful manner when possible. If they cannot be solved, people act to reduce the harms they inflict and minimize damage in their aftermath. Adaptive solutions are many in number and diverse in nature. They require methods for monitoring partners, poachers, allies, and enemies. They require continuous assessments of your own mate value and the esteem in which others hold you. The underlying psychology of conflict management activates behaviors that range from vigilance to violence. And sexual jealousy, once considered by psychologists to be pathological or a character defect, is in fact a supremely important emotion motivating mate-retention solutions. Adaptive in the evolutionary sense of leading to greater survival and reproductive success, of course, does not mean morally good. As we will see, male sexual jealousy is a leading cause of inflicting physical and sexual violence on women.

THE MOST DANGEROUS EMOTION

Most scientific researchers do not consider jealousy to be a "basic emotion." Their argument is that jealousy contains no distinctive facial expression, in contrast to such emotions as fear, rage, and disgust. From an evolutionary perspective, however, carrying a distinct facial expression is important only for emotions that have evolved to signal internal states to others. Anger expressions, for example, send a signal to others to recalibrate their welfare trade-off ratios with respect to the signaler, as we saw in Chapter 3. Jealousy, in contrast, serves adaptive functions other than signaling — it alerts the jealous person to mate poachers, to cues to a partner's infidelity, and to the potentially catastrophic costs of a partner terminating the relationship. Importantly, jealousy provides the internal motivation to take action to ward off present and impending relationship dangers.

Jealousy comes online psychologically when there is a risk to a valued social relationship. When the valued relationship is a romantic one, threats can come from the outside in the form of interested and more desirable rivals. Threats also can come from the inside, from the dual dangers of a partner's infidelity or abandonment. Even if one's partner is 100 percent faithful and shows no signs of straying, jealousy can be triggered by a mate-value discrepancy, which can open up when a partner gets demoted at work, gets fired, loses social status, or suffers a debilitating illness. In these cases, dangers from discrepant desirability may not come immediately; most people do not leave a partner who suffers a temporary setback. Persistent discrepancies, however, loom like ominous clouds on the horizon of the relationship, portending storms if they last too long.

Historically, the reproductive harms of a partner's infidelity or abandonment have been similar for men and women in some respects, so the adaptive functions of jealousy are more or less the same. Sexually similar psychological features include the activation of jealousy when there are genuine dangers from interested and more desirable rivals, observable evidence of sexual or emotional betrayal, being more invested than one's partner in the relationship, and being lower in desirability than one's partner. In both sexes, jealousy serves a motivational function — spurring action aimed at warding off threats, increasing one's mate value to close the gap, inducing the partner to invest more heavily, and ultimately performing acts to retain the partner. In both men and women, failures at mate retention carry the costs of losing valuable reproductively relevant resources — assets that may be difficult or impossible to replace.

In some respects, however, the historical harms of a partner's infidelity or desertion have been asymmetrical for men and women. One asymmetry centers on a fundamental fact about human reproductive biology — fertilization and implantation of the conceptus occur internally within women, not within men. This sexual asymmetry creates a profound sex difference in certainty of biological parentage. Women are 100 percent certain that newborns who emerge from their own bodies are their own. Men can never be sure: "Mama's baby, Papa's maybe." To compound this problem, men risk investing a decade or two in a rival's child in the mistaken belief that the offspring is their own. Consequently, evolutionists have proposed that men's sexual jealousy will focus very heavily on the *sexual* aspects of an infidelity. It is precisely a sexual betrayal — when his partner has intercourse with another man — that threatens to compromise his certainty in paternity and the cascading costs that follow from it.

Although women's maternity certainty is not compromised by their partner's sexual infidelity, women face another key problem — the potential loss of their partner's time, energy, attention, investment, resources, protection, and parenting, all of which could get diverted to a rival woman and her children. Consequently, evolutionists have hypothesized that a woman's jealousy will focus intensely on cues to emotional infidelity, such as whether her partner seems to have fallen in love with another woman. Emotional infidelity is a paramount cue to the loss of her partner's reproductively relevant resources.

Both sexes typically get extremely upset at cues to both sexual and emotional infidelity, as they should — people sometimes fall in love with those with whom they have sex and conversely have sex with those with whom they become emotionally involved. So researchers asked: What would upset you more: (A) Imagining your partner having sexual intercourse with someone else? Or (B) imagining your partner falling in love with someone else? This method revealed large sex differences, with the majority of men, roughly 60 percent, selecting sexual infidelity (A) and most women, roughly 85 percent, selecting emotional infidelity (B).

Different methods confirm sex differences in the psychology of jealousy. Imagine your partner enjoying passionate sex with another *and* becoming deeply emotionally involved with them. *Which aspect* of the infidelity would upset you more? Again, the sexes differed in the predicted ways by about 40 percent — a large difference by social science standards. These sex differences in jealousy have now been replicated in Brazil, Chile, China, England, Ireland, Korea, Norway, Romania, Spain, and Sweden.[1]

The field of psychology is sometimes criticized for studying WEIRD samples — an acronym for Western, educated, industrial, rich, and democratic. That criticism does not apply to the study of sex differences in jealousy. Led by evolutionary anthropology professor Brooke Scelza, a large research team administered the same test to eleven different populations, mostly small traditional foraging societies that are decidedly non-WEIRD. These included the Himba of Namibia, the Mayanga of Nicaragua, the Hadza of Tanzania, the Karo Batak of Indonesia, the Shuar of Ecuador, the Yasawa of Fiji, and the Tsimane of Bolivia.[2] Scelza and her research team found strong support for the universality of the sex difference — men more than women in every culture indicated greater upset about a partner's sexual than emotional infidelity, precisely as predicted. One Himba man explained his reaction this way: "When a man has sex with your wife, it means he eats your cattle, your things," referring to the resources that must now be given to the children from his wife's affairs.[3] In short, Himba men want to avoid diverting their resources to a rival man's children — a cost incurred due to his wife's sexual intercourse with another man.

Professor Scelza, however, found two interesting empirical twists in her eleven-culture study. First, although the predicted sex difference in jealousy proved universal, women in more traditional cultures were more likely than WEIRD urban women to find the sexual infidelity scenario distressing. She speculates that in more traditional non-WEIRD cultures, a man's sex with a woman outside of his marriage provides a powerful indicator that he will not be investing in his main partner and their existing children, but rather diverting vital resources to his sexual affair partner. Consequently, women in these cultures become especially distressed about a man's sexual infidelity. One Himba woman explained it this way: "For the culture it's fine [referring to the Himba's liberal sexual norms] but for me I don't like it when he goes off to have sex with someone else because he could leave me and never come back."[4] This interpretation is supported by another woman, who explained why she chose sexual infidelity as more upsetting: "If he has sex with someone else he might drop me."[5]

Scelza also discovered that cultures differ dramatically in how much parental investment men typically devote to their children. In cultures with heavy male parental investment, men showed an even stronger endorsement of sexual infidelity as more distressing than emotional infidelity; more than 90 percent in those cultures, such as the Himba, chose sexual infidelity as more jealousy inducing. The more investment men make, the more important it becomes for them to ensure that they are the actual genetic fathers, at least from an evolutionary perspective. In short, cultures vary in how jealous men get about a partner's sexual infidelity, but it's not random or arbitrary cultural variability. It is theoretically predictable variability based on how much men invest in their children, which in turn corresponds to the costs they would incur by investing in a child who might just be their rival's.

Sex-differentiated mating psychology also predicts sex differences in jealous actions taken when an infidelity is discovered. In studies of verbal interrogations upon discovery that some sort of infidelity has occurred, men are more likely to say "Did you fuck him?" whereas women are more likely to say "Do you love her?"[6] Women are more likely to forgive a partner's sexual infidelity; men are more likely to forgive an emotional infidelity if it was not accompanied by sexual intercourse with a rival man.

The green-eyed monster of jealousy shows several other sex-differentiated psychological elements. One centers on characteristics of rivals and how large a threat they pose. Men become more jealous when an interested rival exceeds them in job prospects, financial assets, and physical strength.[7] Women's jealousy is more activated than men's when a rival woman is more attractive or has a more sexually appealing body. These sex differences are robust not just in American samples, but also in South Korea and in the Netherlands, one of the most sexually egalitarian countries in the world. Because resources and physical attractiveness contribute somewhat different weights to women's and men's overall desirability, the jealousy each sex experiences is partly a function of whether the interested rivals surpass them on those qualities.

Encountering rivals higher in mate value motivates problem-solving action. Women and men disparage their rivals, for example, with verbal barbs that strike some as vicious. Women more than men describe their mating rivals as unattractive, lopsided, fat, ugly, having heavy thighs, having no shape, and having a "flabby ass" or "no ass." Men more than women impugn their rivals' strength, status, car quality, ambition, drive, and job prospects.

Understanding the underlying psychology of sexual jealousy is critical because it is the leading cause of mate-related violence, especially that perpetrated by men. It causes men to sequester their partners, verbally abuse them, and in extreme cases murder them. Male sexual jealousy is the leading cause of the murder of adult women, accounting for between 50 and 70 percent of all such homicides.[8] Police know this. When women are murdered, the prime suspects are boyfriends, husbands, ex-boyfriends, and ex-husbands. Although jealousy sometimes motivates women to murder, only 3 percent of murdered men are killed by their romantic partners or exes, and many of these female-perpetrated homicides are women defending themselves against a jealously violent man. Although most men who get jealous do not unleash lethal violence on their romantic partners, sublethal brutality in the forms of intimate partner violence (IPV), stalking, and partner rape are disturbingly common — topics taken up at greater length in Chapters 5, 6, and 7.

Before delving deeper into the actions people take once the green-eyed monster rears its ugly head, we must ask: How are infidelities discovered to begin with and how does the arms race of sexual conflict surrounding infidelity, mate guarding, and evasion of a partner's mate guarding play out?

VIGILANCE, INFORMATION GATHERING, AND MATE GUARDING

Once jealousy is activated, vigilance is often the first line of defense against relationship threats. It motivates information gathering. Cues to financial infidelity, sexual infidelity, emotional infidelity, or desirability discrepancy all have the power to galvanize vigilance. As an analog, a gazelle who detects the possible presence of a predator becomes alert, watchful, observant, attentive. How close is the threat? How formidable is the threat? From which direction is the threat coming? Effective strategic action depends on knowledge of this vital information. Without this knowledge, the gazelle might mistakenly flee right into the lethal fangs of the predator. So it is with dangers to relationships. Strategic solutions require deep knowledge of the source, proximity, and magnitude of the threat. Who is the potential mate poacher and what is their mate value? How committed is my partner to me? Why does my partner have a second cell phone? The first step toward effective solutions requires gathering information.

Adaptive solutions also rest on the trail of cues one's lover may have left behind—the clichéd lipstick on the collar. Thinking back, why did my partner come home so late last Friday night? Was he really at work like he claimed? Why is our joint bank account balance so low? What are those mysterious charges on our credit cards? Why does he blank his computer screen whenever I walk by?

Our studies discovered the following actions that people use to gather information when psychological vigilance is on high alert: he called her at unexpected times to see whom she was with; she called him to make sure he was where he said he would be; he had his friends check up on her; she snooped through his personal belongings; he questioned her about where she had been and what she had done when they were apart; she dropped by unexpectedly to see what he was doing; he read her personal mail; she stayed close to him while they were at a party; at a party, he did not let her out of his sight.[9]

In our studies of newlywed couples, we found that women and men were equally likely to perform vigilant actions. Interesting gender differences emerged, however, in which men and which women ramped up their vigilance. Men who rated their wives as highly physically attractive were especially likely to escalate the vigilant forms of mate guarding. They also gave their attractive wives more gifts and jewelry, took pains to enhance their own physical appearance, and acted possessive through verbal signals. For example, a vigilant man might introduce his partner to everyone as "my wife" and drape his arm around her when other men are around. Yes, men have a long and sordid history of treating women as possessions.[10] Men married to attractive women also devote more effort to fending off mate poachers. They are more likely to stare coldly at other guys who direct their gazes toward their wives, confront guys who seem to be making passes at their wives, and tell other guys directly to "stay away" from their wives. Some even pick fights with other guys. A few resort to physical violence, such as punching a rival or keying the paint on his car.

These research findings may be part of common folklore. Indeed, they were well captured by the lyrics to the Dr. Hook song "When You're in Love with a Beautiful Woman."[11] To paraphrase, the song bemoans the anguish a man feels because everybody wants her and everyone wants to take her home. The man yearns to trust her but gets suspicious when there are anonymous hang-up phone calls. He attentively watches her eyes and must be on the lookout for lies. Importantly, men in love with beautiful women have to watch their friends. Friends, as research has shown, sometimes turn into mate poachers.[12] Friends are often in close proximity, share similar interests, have deep knowledge of any conflicts the couple may be experiencing, and are better positioned to implement a mate-poaching strategy. Friends can become rivals, widening the wedge of a couple's conflict until there is enough space to enter the gap.

Men engage in similar forms of mate guarding when their brides are young and when there exists an attractiveness discrepancy — when the man views his wife as more attractive than she views him. For these couples, men deploy greater vigilance; they monopolize more of their wives' time; they are more likely to tear other men down with verbal barbs; and they are more likely to threaten with violence both their partners and men who seem attracted to their partners.

In contrast, the man's physical attractiveness, his age, and attractiveness discrepancies do not seem to influence women's vigilant action in these directions. Indeed, women married to physically attractive men appear to do slightly less mate guarding overall. The man's salary and the effort he devotes to navigating his way up the status hierarchy, however, have a large impact. Women married to higher-income men display more vigilant guarding and put more effort into enhancing their own physical attractiveness through makeup and clothing, although this may simply reflect having more money to spend on these items. These women are also more likely to act submissively as a mate-retention tactic. In short, although both women and men devote equal effort to mate guarding, the triggers of more intense efforts differ. For men, the partner's attractiveness and relative youth drive their efforts up. For women, the partner's income and status striving motivate their efforts.

The personality traits of mate guarders also influence forms of mate guarding. Men high in Dark Triad traits are more likely to engage in high-level vigilance, snooping through their partner's mail and tracking their movements.[13] They intentionally evoke jealousy in their partner by talking with other women, flirting with other women, and even going out on dates with other women. Despite their own flirtatious behavior, high-scoring Dark Triad men come down hard on their partner for flirting — yelling at her, threatening never to talk to her again, and even hitting her. They use emotional manipulation, such as pretending to be angry and making their partner feel guilty when she innocently interacts with other men. High-scoring Dark Triad men are also more likely to glare at other men who are talking to their mate and threaten them with physical violence — a topic we take up in greater detail in Chapter 5. In short, high-scoring Dark Triad men are more vigilant, deceptive, manipulative, emotionally exploitative, and physically threatening in their mate-guarding tactics compared to men who score low in these traits.

WOMEN'S RESISTANCE TO MATE GUARDING

Some men start out as Prince Charming and turn dangerously violent over time. Most women, however, are keenly aware of their partner's jealousy and may take pains not to trigger it or take steps to evade the mate guarding it motivates. The arms race brought about by sexual conflict predicts that men's mate-guarding adaptations, if they are recurrently harmful to the women who are guarded, will favor adaptations in women to resist men's mate guarding. Excessive mate guarding inflicts costs on a woman in several ways. It monopolizes her time, energy, and resources, preventing their allocation to dealing with other important challenges such as investing in career or kin. It interferes with her ability to form backup mates, blocking her ability to cultivate mate insurance should the current mateship turn south. It prevents her from having an affair, which can have important benefits, as noted in Chapter 3. And it prevents her from creating opportunities to mate switch, even when trading up would be advantageous. In short, a man's mate guarding interferes with the woman's ability to exercise freedom to allocate her time as she sees fit and blocks a cardinal feature of women's mating strategy — preferential mate choice. All these harms point to the evolution of tactics to resist mate guarding.

Evolutionary psychologist Alita Cousins and her colleagues have discovered six tactics women use to evade the clutches of men who are overzealous in their guarding efforts.[14] *Covert tactics* include: "I hid stuff from my partner so he wouldn't find it"; "I had other men call when I knew my partner would not be around"; "I avoided situations where I knew my partner might be able to check up on me"; "I was sneaky about flirting with other men so my partner would not find out"; "I told my partner that I was 'going out with friends' when I was really going out with other men"; "I didn't tell my partner that I was going to a party where there would be a lot of attractive men."

A second tactic Cousins discovered involves *avoiding public displays of affection* (PDAs) — refusing to hold hands with the partner in public, not allowing the man to kiss her in public, and even not allowing him to put his arm around her in public. These public displays are well-documented forms of mate guarding.[15] A woman who successfully eludes an overly possessive man prevents him from sending signals to others that his partner is already in a committed mateship. It also allows her freedom to beckon other men with a friendly smile or eye contact that lingers a split second longer. In short, avoiding the man's PDA keeps her mating options open.

A third tactic involves *attempts to suppress the man's aggression toward rival men*. For example, "I told my partner not to yell at other men who looked at me, flirted with me, or made a pass at me." Repelling potential mate poachers is a common male mate-guarding tactic. Some women attempt to prevent their partner from keeping other men at bay. Tamping down a man's belligerence toward other men has the added benefit of avoiding the public embarrassment a woman might experience as a result of his overly intense mate guarding. Suppressing her partner's overly zealous mate guarding signals to others that she is not partnered with an insecure man low in mate value, which would in turn reflect on her own perceived mate value.

Alita Cousins calls the fourth tactic the *covert high-tech strategy*. This includes erasing text messages from phones or social media accounts, changing user names and passwords frequently to prevent the partner from snooping, and explicitly telling other men not to contact her via computer because her partner might find out. Just as people mate guard through technology, their partners attempt to circumvent these efforts by deleting technological traces.

A related tactic involves *avoiding partner's contact*. Women resist their partner's mate guarding by ignoring phone calls and text messages or simply by turning off their cell phones during specific times in order to prevent receiving them to begin with. Just as men use calls and texts to check up on their partner and monitor her movements, women find ways to evade these forms of guarding.

The final tactic, called *resisting control*, involves expressing anger at the partner for being too controlling, fighting with the partner to preserve independence and freedom of socializing, and threatening to break up with a partner who is too smothering. Overly forceful mate guarding hampers a woman's freedom of movement and causes her to have to walk on eggshells for fear of provoking a man's jealousy.

Which women are most likely to deploy this collection of resistance tactics? Dr. Cousins discovered several predictors. Women mated with men who are hypercontrolling is one such predictor. The more controlling the man, the costlier it is for the woman — creating a more profound problem that warrants more urgent effort to solve. Here is one woman's account:

> I was living with an intimidating and controlling man. My refuge was Starbucks. Saying I needed decent coffee, I would regularly escape to the Starbucks down the street and happily text with my back-up boyfriends, being careful to delete all

the texts before I went home. However, when my birthday rolled around, my plan was foiled when he gave me a very expensive cappuccino machine as a gift! Not to be undone, I capitalized on the fact that he loved my cooking. In the afternoons, just before dinner I would conveniently be missing some important ingredient. I would then meet my side-guy in the parking garage of the grocery store! Blacked out windows on my luxury car hid my "crimes." We eventually broke up and I'm happily married to a much less controlling guy.

Which men are most likely to be controlling? Those who are lower in mate value. These men ramp up heavy mate guarding because they believe they have lucked out in attracting a desirable woman and believe that she will be difficult or impossible to replace.

Some men are hypercontrolling because they cannot bestow benefits due to a lack of critical resources — precisely the men whom women are most motivated to leave — and so resort to cost-inflicting mate-guarding tactics. Indeed, women who perceive themselves as more attractive than their partner are more likely to deliberately evade their partner's mate guarding.[16]

Another predictor of evasion tactics is how interested women are in casual sex — how high they are on sociosexual orientation (as measured by the Sociosexual Orientation Inventory, SOI). High-scoring SOI women are precisely those most motivated to stray sexually. Having sex with a man outside of the primary mateship usually requires evading men's monitoring. If a woman has no temptation to partake in extra-pair sex or if she has no desire to leave the relationship, the man can relax his mate-retention efforts. Just as men escalate mate guarding when they detect an increased probability of infidelity, women seeking sex on the side escalate their evasion tactics.

How invested women are in their partners is a third predictor of evasion tactics. Lack of investment predicts breakups, but some do not like to end a relationship before they've cultivated another. This strategy is sometimes called *monkey branching*, named after our tree-dwelling primate cousins who swing from branch to branch but don't let go of one until they grasp another. Women who are not strongly invested in their current mateship are most likely to be looking to mate switch, and monkey branching usually requires evasion to implement.

Having an *avoidant attachment style* is a fourth predictor of women's efforts to elude a partner's mate guarding. Those who score high in avoidant attachment tend to struggle with issues of intimacy, tend not to trust romantic partners, and sometimes sabotage their close relationships to prevent getting too close. Attachment-avoidant women dodge public displays of affection by their primary partner and covertly conceal their interactions with other men. Securely attached women feel less need to elude a partner's mate guarding; they welcome intimacy with their primary partner.

We have been discussing women's evasion of men's mate guarding rather than the reverse for a reason — literally no scientific studies have yet been conducted on men's resistance to women's mate guarding! Because women mate guard just as much as men, though, we can speculate that men deploy many of the same evasion tactics — deleting electronic traces of contact they have with other women, turning off cell phones at critical times, avoiding public displays of affection, and so on. And men likely evade for the same reasons as women — when they are less invested in the relationship, when they seek sex on the side, when they have an avoidant attachment style, or when they want to trade up in the mating market. Although we currently lack scientific findings, there is no scientific reason to expect men to differ from women in the tactics they use to avoid and evade their partner's mate guarding.

COPING WITH RELATIONSHIP CONFLICT

COPING WITH MATE-VALUE DISCREPANCIES

Differences in desirability create a bewildering array of problems for both partners. The more desirable partner may become dissatisfied, creating temptations to cheat, to trade up, or to abandon. The less desirable partner becomes more vulnerable to their partner's infidelities and defections, triggering jealousy and intense mate guarding. The more desirable partner, in turn, must then cope with the burdens inflicted by their partner's jealousy and monitoring. Potential mate poachers pick up on these vulnerabilities. Some seek to widen the wedge with verbal influence tactics such as "He's not good enough for you" or "She doesn't appreciate you" or "You deserve someone better." The emotional toll takes away time, effort, and resources from solving other adaptive problems. Coping with mate-value discrepancies can be taxing.

A key coping strategy requires calculating *welfare trade-off ratios* (WTRs), a concept introduced in Chapter 3 as the ratio of how much value someone else places on your welfare relative to their own welfare.[17] Every relationship requires daily decisions that affect both parties. Should I clean the dirty dishes or leave them for my partner? Should I deal with the crying baby or pretend that I'm asleep? Should I gratify my partner's sexual desires even though I'm not in the mood, or should I tell him I have a headache? Should I use my pay raise to buy a new set of wheels for myself or give my partner a pricey present? These decisions depend in part on how much value you place on your partner's welfare compared to your own. The higher the WTR, the more value you place on your partner's happiness relative to your own. Low WTRs are selfishly skewed, those in which your partner's welfare has a low priority relative to your own.

When making decisions that affect both individuals, WTR is a key determinant. It influences decisions small and large. Doing the dirty dishes versus leaving them for one's partner may be small in the grand scheme of things. Diverting financial resources to an affair partner rather than to one's primary partner is more consequential. Even small daily decisions, however, can accumulate over time to large effect. Small irritants, the daily hassles of life, add up like the proverbial water torture — the first drop that falls on your scalp seems inconsequential; the thousandth drop drives you insane.

It is greatly beneficial to have a partner who does not have a selfishly skewed WTR. Selfishness is one of the hallmarks of narcissism. The narcissist's desires will always come first, a partner's a distant second. WTRs, however, are not permanently fixed. They can be recalibrated. And there is evidence that people have evolved adaptations specifically designed to recalibrate the WTRs of others. This is the core of the *recalibration theory of anger*.[18] Although many view anger as a "negative emotion," evolutionary psychologist Aaron Sell and colleagues propose that expressing anger toward someone has a very specific function — to cause the person toward whom anger is directed to value the angry person more highly.

The recalibration theory of anger applies powerfully to psychological skirmishes within mateships. Recall from Chapter 3 the woman who got upset about her boyfriend talking on the phone with another woman and sharing humor and emotional intimacies with the rival. Her response: "I yelled at him when he got off and refused to have sex that night." Her use of sexual withdrawal to punish is obvious. Her expression of anger, however, has a deeper psychological logic. It sends a signal that her boyfriend's WTR with respect to her is too low. If the boyfriend wants to retain the mateship, he will increase his WTR, curtail psychologically intimate interactions with other women, and channel his emotional energies toward her. Her anger signals that his WTR toward her fails to meet her expectations. He does not value her as much as she believes she is worth.

If he fails to alter his WTR, his partner will lower hers toward him. This sends the relationship on a downward spiral in which each partner places less and less value on the other. That is, if you ratchet down the value you place on me, then you become less valuable to me, so I will ratchet down the value I place on you. Consequently, I become less valuable to you, and the downward spiral continues. The underlying logic is that WTRs are mutually interdependent. If you value me, granting me a high WTR, that influences my WTR toward you. You become more valuable to me precisely because you cherish me. And because the value I place on you has increased, my worth to you has increased. WTRs within mateships, in short, are mutually reinforcing. They can spiral up through positive feedback loops. Or they can spiral down.

Although anger is one key emotion expressed in mateships to alter a partner's WTR, other emotions also play a critical strategic role. One coping strategy used by the less desirable partner is to *intentionally evoke jealousy* in the other. Consider the case of a couple attending a party together. Evoking jealousy can be as simple for the lower-mate-value woman as smiling at another man. The smile causes the man to perceive that she might be sexually interested in him. He then approaches her, displays interest in her, and starts to flirt with her. Her romantic partner witnesses interest from the rival. It evokes his jealousy. The attention she gets from the potential poacher increases her regular partner's perceptions of her desirability. No one wants a mate whom no one else wants, and the volume of attention from potential rivals provides an accurate indicator of her mate value. Evoking jealousy, like displaying anger, causes WTR recalibration, at least if implemented skillfully.

Although both sexes report intentionally evoking jealousy, women are somewhat more likely than men to use this strategy.[19] Importantly, the women most likely to use it are those who believe that their partner is less invested in them than they are in their partner — a cue to mate-value discrepancy. A full 50 percent of women who view themselves as more committed than their partner report intentionally evoking jealousy.[20] In contrast, only 26 percent of women who are equally or less committed deploy this tactic.

Evoking jealousy can be a dangerous tactic because jealousy is a perilous passion, the leading cause of partner violence and homicide.[21] This is where plausible deniability enters the picture. A skillful and timely smile, especially if undetected by her partner, can trigger sexual interest from the other man and sexual jealousy in her partner. The two men are like puppets, their evolved mating programs manipulated by a mere smile. She cannot be faulted for having the social graces of simply being friendly. They are at a party after all. When successful, she has recalibrated her partner's WTR, causing him to value her welfare more than he did before. This may sound like game playing, and perhaps it can be viewed that way. On the other hand, mate guarding is serious business. In long-term relationships, partners get complacent and take each other for granted. Periodically recalibrating your partner's WTR can be an important corrective.

Two key predictors of intentionally evoking jealousy are mate-value discrepancies and Dark Triad traits. Our study of married couples found that the lower-mate-value partner is most likely to evoke jealousy by interacting with potential alternatives in a partner's presence, smiling at them and casually touching them on the arm or shoulder. Flirting works. Metaphorical steam comes out of their partner's ears. High-level Dark Triad individuals, both women and men, are especially likely to intentionally evoke jealousy. When effective, it reduces perceived mate-value discrepancies.

Anger and jealousy are not the only emotions used to change a partner's WTR. Another is *forgiveness*. It signals to the partner the option of reconciliation after a relationship violation. "If you treat me poorly, I might temporarily value you less in order to bargain for better treatment."[22] If the expression of anger causes a partner to value them as much as they expect, then forgiveness is in order. Forgiveness signals conflict resolution, a possible end to the internecine skirmish. Importantly, forgiveness resets both WTRs. It creates an upward spiral whereby you value your partner more because they value you more, and your partner in turn values you even more because you value them more.

In sum, anger, jealousy, and forgiveness are emotions used as levers of power in recalibrating one's own and one's partner's WTRs. They are tactical efforts aimed at closing the gap created by perceived mate-value discrepancies.

THE SERIAL-MATING SOLUTION

Serial mating offers a different solution to problems that beset many mateships. Many Americans and Europeans grow up believing in finding the "one and only," a soul mate, one's true love. Nearly everyone who marries believes it will last. Nearly half of them will be wrong. Conflicts are all but inevitable. In my many studies of dating and married couples, I've never witnessed a mateship entirely lacking in conflicts. Sometimes they can be resolved within. Sometimes people choose to live with conflict, coping as best they can. Sometimes they live lives of quiet desperation. Sometimes they call it quits.

The serial-mating solution requires cognitive reframing. Consider friendships as an analog. As people go through life, they make new friends and sometimes discard old ones. We have childhood friends, middle school friends, and college friends. At work, we form other friendships. And for specialized activities such as bonding over fantasy football, love of puppies, or a penchant for shopping, different friends enter our lives. There are two key points to make about friendship. First, we do not require each individual friend to meet all of our friendship needs and wants. My tennis friends help me to fulfill my athletic competition needs and we delight in analyzing the tactical minutiae of exciting matches. We watch with keen interest as a rapidly rising tennis upstart threatens to depose the reigning champion. My academic friends provide other benefits — exploring novel intellectual terrain, discussing the replication crisis in psychology, identifying scientific unknowns and launching new scientific experiments to explore them. I don't expect a friend from one group to fill a gap in the other.

Second, I rarely expect all friendships to last a lifetime. It's great if they do, of course, and I have a few. But you and your BFF from high school might grow apart when you enter college. Nor are your college friends necessarily perfect allies when you launch your professional career. We don't consider friendships that gradually dissolve to be failures; they might be perfect for the right time and place in your life. We don't judge them by the single criterion of whether they last forever.

Framed in this way, serial mating offers a powerful solution to some mating challenges. Just as we outgrow our friends, we sometimes outgrow our mates. A lover perfect in high school may not make the grade after you enter college and expand your intellectual and cultural horizons. Your hometown boyfriend or girlfriend may seem like a rube to your more cosmopolitan companions.

If trust has been broken, serial mating allows you to start anew by switching to a more trustworthy mate. If you've been whipsawed by your partner's volatile moods, you can upgrade to a more emotionally stable mate. If your partner has a violent temper, you can switch to one with a more agreeable disposition. If your partner has a permanently selfish WTR with respect to you, you can search for someone who values you more. If your mate value has risen, you can trade up to a more desirable partner.

Serial mating has many benefits, foremost of which is jettisoning someone whose jealousy has gotten dangerously out of control and whose mate guarding proves stifling. It offers the prospect of correcting some of the mistakes you made in previous mateships. It often brings with it a renewed flush of sexual excitement. The new relationship brings out fresh facets of your personality, some you never realized resided within. It exposes you to novel experiences and different social networks. Serial mating, in short, provides a bounty of benefits, a solution to the many mating problems that plague people who try to grind it out with the tedium of the long term when the relationship has grown stale or turned toxic.

Pursuing this strategy, however, comes with costs. One cost is uncertainty about future prospects on the mating market. Will there be better mates out there? Will they find me desirable? Humans have evolved a strategy for reducing the uncertainty, as we saw in Chapter 3 — they cultivate backup mates and sometimes start an affair before making the transition. They monkey branch. Although some views of sexual morality frown on this solution, it has tremendous adaptive advantages. Mate insurance dramatically reduces uncertainty. Even if the backups or affair partners turn out not to be enduring over the long run, they provide a smoother transition back into the mating pool than jumping into the cold water alone.

Individuals typically reenter the mating market with "baggage" from previous mateships. One acquaintance, for example, started dating after separating from her husband. She had two young children — usually considered a cost, not a benefit, on the mating market. Her economic resources were sparse, since she had been out of the workforce for a half dozen years and had few marketable skills. Moreover, she was still entangled with her ex in a bitter divorce, battling over custody, alimony, and child support. Although she succeeded in re-mating, she had to settle for a man considerably less financially successful and emotionally stable than her husband. Her new mate also turned out to have a drinking problem. After many drunken episodes that triggered verbal and physical abuse, after the police had been called half a dozen times, after she secured a restraining order on her new mate, she finally called it quits. She decided to go it alone for the time being. Her serial-mating strategy did not turn out well, given the costs she carried into the mating market, which lowered her ability to attract a more desirable new partner.

Not all serial maters come with such a heavy load, but most have burdens. Some have exes who still cling, and in some cases stalk, interfering with efforts to re-mate. Some have crushing financial obligations — credit card debts, loans, car payments. Women are especially likely to suffer economic hardship after a marriage dissolves.[23] Compared to men, women experience a greater loss of household income, are more likely to single parent, and are more prone to plunge below the poverty line. Both men and women, however, carry psychological scars — damaged self-esteem from a partner's verbal abuse, resentment about an ex's philandering, or bruising from a bitter breakup. Time, as the cliché goes, does often heal these wounds. Resentments usually fade. The glow of a new love restores one's sense of self-worth. And sex can be an emotional salve, a means for restoring self-confidence after a breakup. As one woman noted, "The best way to get over one man is to get under another."[24]

Another strategy is to offer assets that you uniquely can provide.

THE IRREPLACEABILITY SOLUTION

Imagine living in a small town containing five bakers and one locksmith. Both kinds of professionals are equally valuable. You need bread to eat and locks for your house. Despite being equally valuable, one is more replaceable than the other. If one baker moves out of town, you have four others to rely on. If the one locksmith moves, you are out of luck and your house is vulnerable. The locksmith is irreplaceable.

By analogy, becoming irreplaceable offers a solution to several problems of mating, the most important being loss through abandonment. There are three potential tactics for becoming an irreplaceable mate.[25] The first and most obvious is one already discussed — maintaining your mate value and keeping mate-value discrepancies to a minimum. A partner who is higher in overall desirability than you are, by definition, will be more difficult to replace. In three separate studies of romantic couples, we found that people with partners who were the same or higher in desirability than themselves were indeed more satisfied with their relationships than those whose partners were lower in mate value.[26] People also devote more effort to retaining same- or higher-value partners; they are more difficult to replace.

Although it feels great to have a partner at the upper edge of your desirability range, this solution can create problems. First, those higher in mate value than you are will be more difficult to attract to begin with. Second, even if you succeed, your higher-value partner will be more likely to abandon you, which is undoubtedly why people in this situation devote more effort to mate retention. Third, even if abandonment does not occur, the higher-value partner is more likely to cheat. Although this is not inevitable, the more desirable partner typically feels more entitled to sex on the side or an extra-pair affair.

So this raises a second path for becoming irreplaceable — having a partner who cannot replace you because of a shallow pool of viable alternative mates in the local mating market. Stated differently, even if you do not fulfill all of your partner's desires, if you do so better than those in the extant mating pool, you will become irreplaceable. The key here is not the discrepancy between you and your mate, but rather the discrepancy between you and the pool of alternative partners available to your mate. In our studies of romantic couples, we found that even when someone was considered more desirable than their partner, the higher-mate-value partner's satisfaction with the relationship remained quite high as long as their partner was more desirable than alternatives in the mating pool.[27]

Although this solution to becoming irreplaceable works if the conditions are met, it is difficult to sustain in the modern world of internet mating. Dating apps open up vast pools of potential mates. Living in a small town with a limited mating market allows a 10 to be happily mated with an 8, as long as there exist no other 9s or 10s in town. Living in a cyberworld containing millions of potential mates opens the floodgates to thousands of 9s and 10s. In the cold calculus of relative mate value, if a more desirable potential mate than my current partner is interested and within reach, I may become dissatisfied with my current partner, which may motivate me to switch.

This brings us to the third way to become irreplaceable — fulfilling your partner's individually unique desires, be it in-depth knowledge of Russian literature or a taste for the latest foodie trends. Everyone can make themselves more irreplaceable through judicious partner selection and the intentional cultivation of irreplaceable qualities. Here are some key ways to become irreplaceable:[28]

- Promote a social reputation that highlights your unique or exceptional attributes
- Recognize unique attributes that others value but have difficulty getting from other people
- Cultivate specialized skills that increase irreplaceability
- Preferentially seek out groups that value what you have to offer — groups in which your unique assets will be most appreciated
- Avoid social groups in which your unique attributes are not valued or in which your unique attributes are easily provided by others already present
- Drive off rivals who offer benefits that overlap with those you can provide

The solution works best when both partners have a shared vision for their unique future; when both feel lucky to have found and attracted each other; and when both truly believe that the other is irreplaceable, even if that belief is a slight bit of romantic exaggeration.

For those who do not experience a happy mateship, minimize conflict through coupling up based on similar mate value, fulfill each other's desires, or become irreplaceable, there exists a dark and dangerous suite of cost-inflicting solutions — last-ditch tactics to retain a mate on the verge of leaving. We turn to these more ominous tactics in Chapter 5.

CHAPTER 5

INTIMATE PARTNER VIOLENCE

> Violence against women is never acceptable, never excusable, never tolerable.
>
> — Ban Ki-moon[1]

IN JUNE 2019, IN JACKSON, WYOMING, a thirty-four-year-old man was convicted of multiple counts of domestic violence. The convictions involved the beatings of two ex-girlfriends. One suffered a black eye, a bloody nose, two facial scars, and a broken arm. The other accused him of ramming his knee repeatedly between her legs after he accused her of "screwing someone" while she was out.[2] She ended up in the hospital bleeding from her labia, which had been split in two places.

This case was unusual in several respects — the police investigated and made an arrest; the prosecution secured a conviction from a jury of ten men and two women; and the perpetrator will serve time in jail. In sharp contrast, 75 percent of partner violence incidences are never reported to the police. Of those reported, many are not investigated; of those investigated, many are not charged; of those charged, undoubtedly the more serious ones, many are not convicted; and of those convicted, only about a third do jail time. One study of 517 cases of domestic violence incidents found that fewer than 2 percent of the batterers ended up spending any time in jail.[3] The Wyoming case was unusual in landing in that rare 2 percent. Although the consequences were exceptional, the circumstances surrounding the violence were not — a man's accusations of his partner's sexual infidelity and the temporal proximity to romantic breakup.

As couples careen toward a breakup, some individuals shift tactics from vigilance to violence. Although legal definitions vary from state to state, intimate partner violence (IPV) refers to "a pattern of behavior and tactics used to gain or maintain power and control over a current or former intimate partner that can include physical, sexual, emotional, economic, or psychological abuse or threats of abuse."[4] It is often a last-ditch effort to hold on to a partner who wants to leave. Some people inflict costs when they lack the resources to confer benefits, although some men abuse even if they have resources.

Women in many countries do not have legal protection against IPV. Russian law, for example, has no provisions specifically for intimate partner violence. Nor does it have any procedures in place for imposing restraining orders on men who beat their partners. Russian women have taken to posting selfie photos of their bruised faces and bodies in an effort to draw attention to the fact that at least one in five Russian women suffers violent abuse from the men who claim to love them.[5] These efforts are drawing international attention. A European court ruled in July 2019 that Russia had failed to protect a woman who had been assaulted, stalked, and then kidnapped by her ex-partner. Media attention, one hopes, may prompt Russia to implement laws and attendant penalties.

Within the United States, IPV is illegal in all states, but it is so prevalent that states are often overburdened with cases. Annual rates of violence against women by romantic partners range from 14 percent to 15 percent, and lifetime prevalence rates (the percentage of women who experience IPV at some point in their lives) are estimated to hover around 30 percent in the United States and 27 percent in Canada.[6] These statistics translate into millions who suffer IPV. The crime is so common that federal authorities stepped in to help relieve states of some of the burden. In 1994, the Violence Against Women Act made IPV a federal crime.[7]

There are three general theories, and subtheories within these, that social scientists have advanced to explain IPV: pathology, social learning, and patriarchy. One pathology theory, for example, views IPV as stemming from disorders of attachment, primarily insecure attachment disorder.[8] Men characterized by an insecure attachment style — men who are especially anxious, distrustful of others, and profoundly fearful of partner abandonment — are somewhat more prone to aggress against their loved one. Social learning theory, in contrast, does not view IPV as a pathology, but rather as a learned behavior pattern caused by men's observed "models" in the form of fathers or peers who beat up their wives or girlfriends. According to this theory, observing a father slap or hit a mother provides a role model of how men are supposed to act toward women.

By far the most commonly advocated explanation invokes patriarchy or patriarchal ideologies.[9] According to this view, gender inequality within patriarchal societies is the main culprit. Patriarchal cultures transmit the ideology that men deserve to hold the power and that women should be subordinate to men. Socially inculcated gender roles, taught in childhood and perpetuated through adolescence and adulthood, cause men to slap, punch, or threaten women as tactics to maintain their power and to keep women subordinate to them. Solving the many tragedies of IPV, in this view, requires curing men of their patriarchal beliefs in the short run and overturning patriarchal social structures in the long run.

All these prior theories clearly have elements of truth. At the same time, they do not go deep enough in explaining the root causes of IPV. I will suggest that partner violence has a functional logic. Sexual conflict theory provides a powerful explanatory framework within which to understand IPV and the specific circumstances in which it occurs. This chapter delves into these circumstances in detail.

My explanation will not negate the *observations* of prior theorists. Scholars of all theoretical stripes agree that some men use violence to maintain control over women. This is an incontestable fact. Patriarchy theorists persuasively argue that some forms of institutionalized patriarchy, such as laws that give husbands rights to abuse their wives with impunity, foster men's violence toward women. Nor does my theoretical perspective deny social learning. There is no doubt that humans are adept at learning from others. We attend to and imitate the tactics we perceive to be effective at attaining desired goals. If a boy grows up observing that his mother obeys his father's wishes when he threatens violence, it would not be surprising if the boy remembered the effectiveness of this tactic and used it in future relationships to control women.

An evolutionary perspective, however, helps to explain several known facts about IPV. It offers insights into why only some men resort to violence whereas others recoil with moral horror at the mere thought of it. It helps to explain why IPV is also found in nonpatriarchal cultures such as Sweden, Denmark, Finland, and Iceland. These highly gender-egalitarian countries have lifetime partner violence rates of roughly 30 percent, somewhat higher than those in other European Union countries (around 22 percent) and Australia (around 25 percent).[10] An evolutionary perspective on IPV illuminates why male-to-female violence centers so heavily on *controlling women's sexuality*. And it sheds light on the very specific circumstances, detailed in this chapter, in which men resort to violence as a tactic of gaining control over their romantic partners.

Some episodes of domestic violence are bidirectional, altercations in which women and men both abuse in a physical altercation. In some of these cases, although not all, women are defending themselves against men who are initiating the abuse. The more serious the abuse, however, the larger the gender disparity. Most IPV victims who end up in the hospital are women. Men's violence typically does substantially more damage. Every major city has shelters for battered women. In contrast, one of the country's only shelters for battered men opened in Dallas in the year 2017. Women undoubtedly abuse men, but they generally inflict less damage.

Some battered men are reluctant or too ashamed to seek shelter, of course, and so the actual rates of women battering men may be underreported. And when they are reported, police may be reluctant to act. I know of one case in which a man called the police after being badly battered by his wife, his head bleeding profusely from a blow from a frying pan. Upon arrival, the police discouraged him from reporting the crime, despite his obvious injuries: "If she so much as broke a fingernail, you will likely be arrested, not her." He declined to press charges. Studies confirm that police often trivialize men battered by their mates.[11] Women's violence against men cannot be ignored. Nonetheless, because of the strong gender asymmetry in magnitude of harm from IPV, our discussion will focus heavily on men's use of violence to control and manipulate the women they claim to love.

HOW PARTNER VIOLENCE HIJACKS VICTIM PSYCHOLOGY

From an evolutionary perspective, both women and men are extraordinarily valuable reproductive resources to each other. In the amoral universe of natural and sexual selection, women's bodies and sexual access to their bodies are indispensable assets for men. Losing access historically could put a man on a road to an evolutionary dead end, especially if the lost mate could not be replaced with another. At the same time, men's investments in women and their children, as well as the protection men provide them from exploitation by others, are vital fitness resources for women. Their loss historically could jeopardize a woman and her children, putting them in perilous straits. Consequently, both sexes devote tremendous effort to keeping the commitment of mates they have succeeded in attracting.

Successful mate retention requires influencing a partner's underlying psychological machinery. At the most general level, there are two strategies for influencing that machinery — bestowing benefits and inflicting costs. In an ideal world, everyone would lavish benefits on their mates. When their deepest desires are fulfilled, lovers are highly motivated to stay. Unfortunately, fulfilling desires by benefit bestowal is not always possible. A man might lack the personality, sense of humor, sexual skills, or resources that a woman desires. A woman's low sex drive might render her less motivated to fulfill a partner's sexual yearnings. For both sexes, alternative mates sometimes offer temptations to stray. The appeal of harm infliction as a mate-retention tactic increases when a person has insufficient benefits to bestow.

Proposing that partner violence has adaptive functions, of course, does not mean that it is invariantly expressed. Nor does it imply that IPV should be condoned, excused, or in any way justified. Indeed, knowledge of its functions sheds light on the precise circumstances in which it occurs and hence provides useful information for interventions to reduce its occurrence.

Nor is IPV the primary mate-retention tactic of choice. Violence is often a last-ditch effort to coerce a woman to stay precisely when she is most motivated to leave. Moreover, using violence can be quite costly to the perpetrator. He risks retribution from the victim or her kin. He risks damage to his social reputation. And in most modern Western countries, he risks a restraining order and criminal record. Some states now *require* police to arrest a man when there is evidence of domestic violence, even if the victim refuses to press charges.

An important study by A. J. Figueredo and his colleagues found that, among Sonoran Mexicans, men who lack economic resources are more likely to resort to violence — precisely as predicted based on what we know about women's mate preferences.[12] When a mate-value discrepancy widens, women become tempted to trade up in the mating market. This creates a perfect storm from the perspective of sexual conflict theory. The action that best serves the interests of the woman — leaving a lower-quality man for one of higher quality — conflicts with what is in the best interests of her primary mate. A man lacking the resources to fulfill her desires may resort to violence, threats of violence, or both in a last-ditch desperate attempt to get her to stay. These circumstances do not inevitably lead to violence; they simply increase its likelihood.

Abuse works by hijacking the woman's psychology in several ways. First, it *lowers the victim's self-esteem.* Consider the statement "He calls you names [e.g., bitch, whore] to put you down and make you feel bad." In a study of 8,385 mateships (married or living together), IPV and this type of denigration were tightly linked. Among women who reported no physical violence from their partner, only 3 percent reported that their partner undermined their self-esteem in this way; 22 percent of women who reported nonserious violence had partners who undermined their self-esteem through this form of verbal abuse; and a whopping 48 percent who had been seriously battered had partners who undermined them in this way.[13] As one abused woman noted, "I had no confidence and my self-esteem was at zero. I was told by my abuser that everybody hated me. I was embarrassed, I was ashamed, and I worried I wouldn't be believed so I stayed so much longer than I should have."[14]

Physical abuse often leaves physical marks — bruises, scars, and sometimes broken bones. Because physical appearance is a key component of women's perceived mate value, the damage literally lowers a woman's desirability to other men, at least temporarily, and if repeated, more enduringly. Abuse has the pernicious effect of changing her perception of how she might fare on the mating market. Sometimes a woman comes to believe that she is lucky to have her mate, even if he's abusive. She feels so wretched about herself that she views herself as unworthy of any other man.

Carrying the physical marks of violence inflicts another psychological burden — the emotion of *shame*, a second way abuse hijacks women's psychology. Battered women become profoundly embarrassed about appearing in public, seeing friends, or visiting family. Many abused women feel so ashamed that they conceal the abuse. They cover their bruises with heavy makeup. They wear sunglasses and turtlenecks. The victim's shame further isolates her socially, tipping the balance of power further in the abuser's direction. And if friends uncover what's really behind the makeup and masking, victims of abuse often make excuses and blame themselves: "It was my fault; I shouldn't have provoked him." In sum, abuse evokes shame in the victim; shame leads to self-isolation; and isolation deprives victims of the social support they so desperately need to short-circuit the cycle of abuse and leave the abuser.

Although women are sometimes chastised for concealing abuse, Dr. Daniel Sznycer and his colleagues propose that shame has evolved to solve a critical adaptive problem — to prevent being devalued by others in the group.[15] When a woman is perceived to be less valuable to others and less able to stand up for her own interests, they propose, then others in the group will place less value on her welfare. So shame might motivate abused women to conceal their abuse for a very important adaptive reason — to avoid being devalued by others. Education about the adaptive logic of shame might help abused women escape the vicious cycle of self-blame and may also help others to stop blaming women who desperately try to conceal events that could cause them to be devalued by the group.

Abuse hijacks a woman's psychology in a third way — by *altering her welfare trade-off ratio* with respect to the abuser. In an effort to protect herself from further violence, she comes to prioritize the abuser's wants and needs above her own. Some women describe their own behavior as "walking on eggshells," deferring to their mate lest they provoke his wrath. This WTR shift is sometimes so pronounced that the victim neglects her own children, especially if her current partner is not the father of those children.

A fourth way that abuse hijacks women's psychology centers on the abuser *cutting off the victim's access to resources*. Controlling men sometimes force women to quit their jobs, prevent them from going back to school to acquire education that would qualify them for good jobs, and dole out cash so sparingly that some women barely have enough to buy food for their children. One study of 8,385 mated women found that battered women were five times more likely than non-battered women to agree with the statement "He prevents you from knowing about or having access to the family income, even if you ask."[16] The more numerous the violent episodes a woman reported, the more likely the man was to cut off her access to financial resources. Controlling resources limits a woman's ability to leave an abuser. It took one woman three years of secretly stashing coins and small bills before she amassed just enough to get herself and her four children to safety.[17] This brave woman suffered terror from the threats of her ex in the aftermath of her departure. She went into hiding, but he tracked her down. He shot a gun at her safe house, although fortunately no one was hit. He deliberately cut her and their children off financially, refusing to pay legally mandated child support payments, in an attempt to starve them into returning. She resisted that manipulation. And she imparted a key lesson to her daughter: "Develop your career and always have your independent source of income; never rely on a man for money." That daughter took the lesson to heart and is now a highly successful professor at a prestigious university. Pure economic practicality explains why women who have their own independent sources of income are more likely to leave abusive men. Unfortunately, lack of economic resources also explains why some women at shelters for battered victims return to their abusers — they have the mouths of their children to feed.

A fifth way abusers hijack victim psychology is *gaslighting*. The term stems from the 1944 movie *Gaslight*, starring Ingrid Bergman playing the part of Paula. She is a young bride married to an older man who plans an insidious tactic for manipulating her. He causes the lights, which then were fueled by gas, to flicker at random times. When Paula draws attention to the flickering, her husband insists that the lights are burning steadily. Over time, he continues to undermine her perceptions of reality, ultimately leading her to believe that she can't trust her own eyes and that she must be going insane. Perpetrators of gaslighting typically present victims with false information, contradictory information, or ambiguous information, creating confusion and anxiety. Concrete examples of gaslighting, after presenting confusing or misleading information, include: "Why are you making things up?"; "It's all in your head"; and "You are imagining things."[18] Victims start to distrust their own perceptions and judgments, increasingly relinquishing control over their perception of reality to the gaslighting perpetrator. Gaslighting is a particularly pernicious form of abuse because it can be diabolically difficult to detect. Unlike physical abuse, which is more easily spotted and often leaves bruises and scars as evidence, gaslighting is emotional abuse that often flies under the radar. It typically starts out small and gradually escalates. A woman might detect subtle signals that her husband is cheating, but he convinces her that she's imagining things. The ultimate goal of the gaslighter is to seize control and infiltrate the victim's mind so that she must rely on her perpetrator for perceptions of reality. The gaslighter seeks to zombify the victim's mind.

A sixth way in which men's violence manipulates victims is through *threats of future harms*. These come in several key forms. One is threats of violence. Another involves financial threats, promising penury should the victim disobey him. Still another entails threats of injury to her loved ones — her family, her friends, her children, or even her pets. A Florida man allegedly threatened to kill his girlfriend's pet puppy if she did not return home.[19] She refused, and he allegedly carried out the threat in a jealous rage, leaving the strangled puppy on her doorstep.

At their most extreme, abusers issue death threats. Police record many instances of abusers issuing variants on the refrain "If you ever leave me, I will track you down to the end of the earth and kill you" or "If I can't have you, no one can."[20] Interested readers can discover more about mating motives for murder in my book *The Murderer Next Door*.[21]

Many women succumb to these forms of psychological terrorism. They remain with an unwanted man not because of love or the benefits he provides, but rather to avoid the horrific costs of him carrying out his threats. With a psychological gun to her head, a woman might remain in a relationship to stay alive or to protect her children, even if she yearns to leave. She stays in a psychological prison created by her abuser. In some cases, the gun is literal. In our study of people's memories about when they felt their lives were in danger, one woman reported that her husband slept with a gun under his pillow to back up his death threats.[22]

In sum, abusers use six key tactics that hijack the victim's psychology. Although repugnant, cruel, and morally abhorrent, these tactics often bend the victim to the abusers' will. And abusers typically have very specific goals, to which we now turn.

MEN'S MOTIVES FOR PERPETRATING PARTNER VIOLENCE

Sexual conflict theory specifies the precise problems that those who use IPV attempt to solve, and consequently the specific circumstances that put partners at greatest risk. The first is the problem of *mate poachers*. Mate poaching turns out to be a surprisingly common mating strategy.[23] In American samples, for example, 93 percent of men and 86 percent of women reported having attempted to lure someone out of an existing relationship for a long-term mateship. Similarly, 87 percent of men and 75 percent of women report having attempted to poach for short-term mating goals. Although rates of reported mate poaching vary from culture to culture, the vast majority of individuals have experienced mate poaching — as a mate poacher, as the recipient of mate-poaching attempts, or as the "victim" whose mate someone attempted to lure for a short-term liaison or a long-term mateship. Mate poachers create a serious challenge because they threaten to usurp the resources of the "victim," the primary partner, that were previously inherent in the mateship.

When violence is used, it is sometimes directed at the mate poacher rather than at the mate.[24] Nonetheless, men often direct violence toward their intimate partner when faced with the threat of a mate poacher. Battered women, compared with women in non-abusive relationships, agree with the following statements much more frequently about their intimate partner: "He is jealous and doesn't want you to talk to other men"; "He tries to limit your contact with family and friends"; and "He insists on knowing who you are with and where you are at all times."[25] In short, heavy mate guarding and violence often co-occur, especially when abusers feel threatened by a potential mate poacher.

Even when the violence is solely directed at a partner, it can send a signal indirectly to the mate poachers — "I'm the type of guy who resorts to violence — don't fuck with me or get anywhere near my woman." As evolutionary psychologists Margo Wilson and Martin Daly note, "Men are known by their fellows as 'the sort who can be pushed around' or 'the sort that won't take any shit,' as people whose word means action and people who are full of hot air, as guys whose girlfriends you can chat up with impunity, or as guys you don't want to mess with."[26] Acquiring a reputation for violence can deter potential mate poachers.

The possibility of *sexual infidelity* presents another challenge. Men use an array of tactics to solve this threat, and violence is one tactic in that arsenal. Indeed, the detection or suspicion of infidelity is one of the key predictors of intimate partner violence.[27] In one study, battered women were interviewed and then divided into two groups.[28] One group had been both raped and beaten by their husbands. A second group had been beaten but not raped. These two groups were then compared to a control group of non-victimized women. The women were asked whether they had "ever had sex" with a man other than their husband while living with their husband. Ten percent of the non-victimized women reported having had an affair; 23 percent of the battered women reported having had an affair; and 47 percent of the women who were both battered and raped reported having committed adultery.

These findings, if taken at face value, suggest that female sexual infidelity may dramatically increase a woman's risk of being battered. Causality, of course, cannot be determined from this study. It is possible, for example, that men who batter and sexually assault their wives drive them into the arms of other men. But from an evolutionary perspective, one would expect that the infidelity or the possibility of it comes first. Abusive men endeavor to prevent a partner's sexual infidelity and to deter future infidelities, which jeopardize their paternity certainty and risk the allocation of resources to a rival's child. None of this evolutionary logic, of course, is consciously calculated by abusers.

One of the most disturbing predictors of a man perpetrating partner violence is *when his partner gets pregnant.* On the surface, this finding is puzzling. If men have adaptations that evolved because they facilitated reproductive success, why would a man endanger a fetus carrying their genetic cargo? One answer to this mystery lies with the possibility that the abuser suspects that the woman might be pregnant not from his seed but rather from a rival's. Sexual infidelity can lead to impregnation by a rival — a disastrous outcome in the currency of the regular partner's reproductive success. If the fetus is carried to term, the man risks investing in the offspring of his reproductive competitor. To compound this cost, he loses his partner's parental investment, since her maternal resources would be diverted to the rival's child instead of his own. Pregnancy with another man's child is a distinct, albeit closely related, adaptive problem to that of the woman's sexual infidelity in that the hypothesized function of violence differs in the two cases. In the case of infidelity or infidelity threat, the man's violence is directed at deterring infidelity or future episodes of infidelity. When a woman becomes pregnant with another man's child, in contrast, the hypothesized function of violence is abortion — to terminate the pregnancy and eliminate the fetus of a rival.

Another speculation is that, even if the man is reasonably confident that the incipient child is his, the timing of the pregnancy may not be auspicious because of his limited resources, because of his desire to allocate effort to other tasks such as status attainment, or simply because he does not want to tie up the next two decades of his investments in his current partner and the child she is carrying. Just as women exercise choice to time their pregnancies in propitious circumstances, such as with the right man with the right genes who will be a good dad, men exercise choice for similar reasons. An untimely or unwanted pregnancy upends these critical decisions, and a cruel and selfish man may use morally abhorrent means to regain his perceived prerogative.

Is there any evidence to support the fetus-murder hypothesis? Yes, albeit provisional. The frequency of violent acts toward pregnant mates roughly doubles compared to that directed toward partners who are not pregnant.[29] Furthermore, men who abuse their pregnant partners are more likely to be sexually jealous and perceive that the child might not be their own, even if that perception is delusional. This evidence is circumstantial. A more direct test, however, comes from a study that compared violent and nonviolent couples and found that women abused while pregnant were in fact more likely to be carrying the child of a man other than their current mate.[30]

Another prediction from this hypothesis is that the specific form of violence will be aimed at producing the highest probability of aborting the fetus, such as blows to the woman's abdomen. A study in Nicaragua found precisely that. Half of a sample of pregnant women who were abused had suffered blows directed specifically to their gravid bellies.[31]

Are these acts morally abhorrent? Absolutely. Few things seem more evil than directing abortion-inducing blows to a pregnant woman's belly. Are such reprehensible deeds functional through an evolutionary lens? Possibly. Evolution by selection operates according to the ruthless currency of relative reproductive success. It is indifferent to the suffering of individuals. It is indifferent to our moral evaluations. Just as evolution has created male lions that kill the baby cubs sired by their rivals in order to bring a new female back into estrus, human males may have evolved a circumstance-contingent psychology that inclines them to brutalize incipient offspring sired by rival men.[32] Whether or not this hypothesis turns out to be correct when more tests are conducted, the key point is that an evolutionary lens has heuristic value for predicting the very specific circumstances in which intimate partner violence is likely, and even the particular forms it is likely to take.

Stepchildren pose another key risk of partner violence. A Canadian study found that women with children sired by a previous mate sought the protection of shelters for battered women at an astonishing five times the rate of same-aged women whose children were all the progeny of their current mateship.[33] Unrelated children residing in the home create multiple challenges for intimate relationships.[34] From the perspective of the stepparent, a stepchild typically is viewed as a cost, not a benefit, of the mating relationship. Resources from the stepparent get channeled toward the offspring of rivals. Compounding these costs, the woman's parental resources also get channeled toward her own children, who are the biological children of the new man's reproductive rivals. Escalating these costs even more, the presence of a stepchild may delay a new pregnancy. Breastfeeding, for example, tends to suppress women's ovulation, a natural birth-control and birth-spacing device.[35] So a woman who breastfeeds a man's stepchild has reduced odds of becoming newly pregnant. Even if she is not breastfeeding, the woman may be reluctant to have another child while she has a young child so heavily dependent on her. Natural birth spacing in humans hovers around three to four years. Delayed reproduction, in short, adds another cost in the currency of evolutionary fitness to a man who takes on responsibilities for stepchildren.

Finally, if and when a stepparent succeeds in reproducing with a new mate, his children will be half siblings rather than full siblings with the stepchildren. The decreased genetic relatedness among children residing in the same household intensifies the conflicts of interest among them. Competition for scarce parental resources, sometimes extreme among full siblings, intensifies even more when siblings are less related genetically.

Children of differing genetic relatedness to the two parents can also create resource conflict between the parents. A man might be motivated to withhold resources from the stepchild in favor of his own children, especially in lean times. These propensities may explain why stepparents typically invest fewer resources, such as dollars for college education, in stepchildren than in genetically related children.[36] These inclinations help to explain why physical abuse of stepchildren is many times higher than physical abuse of children residing with both genetic parents.[37] They help to explain why stepchildren leave home a full two years earlier than children residing with both genetic parents. And they help to explain why being a stepchild is the single largest risk factor for getting killed as an infant or young child, far exceeding other contributors such as poverty or low socioeconomic status.

The genetic parent also faces challenges as a consequence of partnering with someone other than the genetic father or mother of the child. A woman, for example, can be torn between two goals inherently in conflict with each other. One goal is securing investment for her child. The second is securing a long-term committed mateship. If a woman's child is perceived as interfering with her new mateship, she may be inclined to withhold resources from the child in order to solidify the mateship. In extreme cases, such as Diane Downs or Susan Smith, the mother may attempt to kill her own children in order to clear the way for a new mateship; although these extremes are exceedingly rare, they reveal the tip of the iceberg of conflicting interests created in stepfamilies.[38]

If all these points are valid, why would a man ever mate with a woman who already has children by another man? Some men are perfectly willing to mate with women who carry the costs of stepchildren if there are offsetting benefits. The main one is that the woman is higher in mate value than he is, all else equal. So a man who is a 6 is willing to mate with a woman who would be an 8 with no children, but her perceived mate value is dragged down by having those children. Once mated to her, though, the man sometimes treats those stepkids more poorly than he does his own children, although of course there are many exceptions. Some men continue to invest in those stepchildren in order to preserve their mateship with their mother, and some men grow genuinely fond of those children and treat them as though they are their own.

These arguments do not imply that there are adaptations specifically designed to kill or even to abuse stepchildren. Stepchildren, unlike lion step-cubs, are rarely killed. Most stepparents invest at least some resources in their stepchildren. Such investment, from an evolutionary perspective, is typically considered to be "mating effort" rather than "parental effort."[39] That is, the true function of this form of investment is to secure access to a mate and their resources, not to increase the survival and success of the stepchild.

Even though we lack knowledge of the specific psychological mechanisms that explain violence toward stepchildren, the occurrence of such violence is clearly explicable from an evolutionary understanding of the conflicts of interest inherent in intimate relationships that involve the presence of stepchildren.[40]

Mate-value discrepancies also can trigger partner violence. One source of discrepancies occurs through errors of selection. An individual may have successfully deceived a prospective mate about their resources or florid sexual history, for example, before committing to the current mateship.[41] The consequence is that the deceiver may be lower in mate value than initially perceived. A second source of discrepancies occurs when a hidden cost does not come to light until after the mateship has been formed. One person might turn out to have financial debts that were concealed. Another might turn out to be secretly in love with their first romantic partner. Either partner might turn out to have extended kin who siphon off resources, or personality dispositions such as emotional instability or narcissism that inflict a heavy "relationship load."[42] Another source of discrepancies occurs when individuals mate while young, prior to establishing an accurate assessment of their own mate value. A teenage girl, for example, may get taken out of the mating market by an older man who secures her commitment before she has an accurate self-assessment of her level of desirability.

Mate-value discrepancies predict intimate partner violence, especially when the woman in a heterosexual relationship emerges as higher in mate value, for three key reasons. First, the partner with the higher mate value is statistically more likely to be sexually unfaithful — a known predictor of abuse.[43] Second, the higher-value individual is more likely to think about trading up in the mating market and so may give off cues to leaving. Third, if the man is lower in mate value, he will have greater difficulty providing resources to the woman, which also increases the odds of infidelity or breaking up. All of these variables may tip the scale toward using violence as a method of mate retention. In support of this hypothesis, there is evidence that those lower in mate value in fact do show more controlling and aggressive behavior toward their partners.[44]

Mate violence can serve at least two related functions in the context of mate-value discrepancies. First, it can directly deter a mate from the temptation to stray or defect.[45] Second, it can reduce the perceived magnitude of the mate-value discrepancy in the eyes of the victim. Being abused verbally, psychologically, physically, or sexually typically lowers an individual's self-esteem, as noted earlier.[46] The abused woman might come to feel that she is unattractive and undesirable and may even be convinced that her abuser is the only man who would have her. As abhorrent as this idea is, mate violence may serve the functions of infidelity deterrence and mate retention by damaging a woman's self-esteem, reducing what she might otherwise perceive as a mate-value discrepancy.

Violence can also *prevent a partner from breaking up* and can *coerce a partner to return to a mateship*. Just under half of all marriages in America end in divorce. Mateship dissolution typically comes with a large loss of the partner's reproductively relevant resources. Consequently, when the net benefits of keeping a partner outweigh the net benefits of alternative options, we expect motivations to prevent a partner from defecting. A partner's abandonment can also damage the social reputation and mate value of the person who is rejected. The discovery that someone was ousted by their previous partner has a negative impact on people's desire to pursue a romantic relationship with them.[47]

Solutions to the adaptive problem of defection, like solutions to many of the adaptive problems we have been discussing, range from elevated vigilance to escalated violence. Those who are unceremoniously rejected use a variety of coping strategies, including physical threats, violence, and stalking.[48]

Unfortunately, violent tactics sometimes work. Some battered women remain in violent relationships. Some return to their abusers even after they have sought help at a shelter. In a study of one hundred women at a shelter for battered women, twenty-seven returned to their partner after he promised that he would change and refrain from violence.[49] An additional seventeen returned as a direct result of threats of further violence if they did not return. As Margo Wilson and Martin Daly note, "A credible threat of violent death can very effectively control people."[50] Another fourteen returned because they had nowhere else to go, and thirteen stated that they returned because of their children. Eight returned because they said they were still in love with the man or felt sorry for him. In short, an astonishing 79 percent of battered women ended up returning to live with their abuser.

Intimate partner violence, of course, does not always succeed in getting a partner to remain in a relationship. It can backfire on the abuser, as some women find avenues for escaping from a cruel mate. Violence may represent a last-ditch tactic to keep a mate who has already decided to leave, suggesting an escalating deployment of desperate tactics of mate retention.[51] Nonetheless, based on existing evidence, we cannot discount the possibility that in some contexts, violence functions to prevent a partner from leaving, giving the abuser some degree of temporary or longer-term access to the partner.

The global COVID-19 pandemic of 2020 proved disastrous for those in abusive mateships and for couples in the process of breaking up. Although during the pandemic many violent crimes such as mugging and murder saw precipitous drops, intimate partner violence, both physical and sexual, spiked by roughly 20 percent. Victims were forced to live within the same constrained spaces as their abusers. Many shelters for battered women closed, cutting off important routes of escape. Violent partners abused with impunity as victims were forced to live without the protective cocoon of their kin and allies. Social distancing prevented many "bodyguards" from rushing to the aid of victims. Those seeking a breakup because of abuse were forced, due to physical and economic constraints, to endure a live-in partner from whom they wished to flee.

Cultural, social, and legal protections are needed to help to short-circuit these violent forms of mate retention. Despite the temporary spike of IPV that occurred during the pandemic lockdown, there are encouraging signs that the overall rates of IPV have been declining over the past few decades.[52]

WHICH MEN ABUSE AND WHICH WOMEN ARE MOST VULNERABLE TO VIOLENCE?

We have outlined the critical circumstances most aligned with partner violence, ranging from mate-value discrepancies to last-ditch efforts to prevent a mate from leaving. But not all men resort to violence in these circumstances. We now know which ones are most prone to employ this tactic — men who are high in Dark Triad traits. A study of 380 Japanese individuals living in Tokyo illustrates this important conclusion.[53] The researcher assessed several types of IPV — direct violence, verbal abuse, sexual abuse, financial controlling behavior, and stalking. Those high in Dark Triad traits were more likely to use all of these tactics. Among the three Dark Triad traits, psychopathy proved to be the strongest predictor.[54] Men high in the psychopathy trait ramp up their mate-retention tactics, including both nonviolent and violent means of control.

These traits show up in the most extreme abusers — men who have been convicted of battering their partners and are required to attend an intervention program designed to curtail future violence. These men showed high-level Dark Triad traits, enough to qualify them as having diagnosable personality disorders, such as narcissistic personality disorder and antisocial personality disorder.[55] Disturbingly, high scores on these same traits within the normal nonclinical spectrum also predict IPV among college students.[56]

Another personality cluster that predicts IPV is borderline personality disorder (BPD).[57] The hallmarks of BPD include a marked fear of being abandoned, extreme swings of mood, explosive anger, unstable self-image, self-destructive behavior, and difficulty in controlling impulses. People with this personality type identify strongly with the following statements: "I am constantly afraid that the people I care about will abandon me or leave me"; "My emotions shift very quickly, and I often experience extreme sadness, anger, and anxiety"; "When I'm feeling insecure in a relationship, I tend to lash out or make impulsive gestures to keep the other person close"; "I would describe most of my romantic relationships as intense, but unstable."[58] What is especially interesting about BPD is that it is not linked with aggression in general; rather, aggression seems to be specific to intimate partners.

A book entitled *I Hate You — Don't Leave Me* captures the psychological essence of why those characterized by BPD are especially violent toward intimates.[59] Because they fear abandonment by someone they love, their inability to control their emotions and impulses leads them to lash out at the people they fear will leave them. Tragically, those with BPD often create a self-fulfilling prophecy. Their aggression and insecurity sometimes drive their partners away, bringing about precisely the abandonment they most dread.

The combination of personality and abuse-predictive circumstances creates a volatile mix. When a man high in Dark Triad traits or with a pivotal personality disorder is paired with a partner who has higher mate value or with social situations such as a lack of economic resources, the presence of mate poachers, or stepchildren in the home, danger lurks and abusers often explode with rage.

If this evolutionary hypothesis is correct — that men have an evolved psychology that functions to attract, control, and retain women high in reproductive value — then we should expect men to allocate more intense mate-retention efforts to women who are young and attractive, two key cues to a woman's overall mate value. Male competition intensifies around young attractive women, whether in the form of hundreds of clicks on internet dating sites like Tinder and Match or in the form of trying to lure a desirable woman out of an existing relationship for sex or romance. Because young attractive women have many mating opportunities, they will be more motivated to leave an unhappy mateship and more likely to trade up in the mating market when they can.

Studies of IPV conclusively show that young wives are indeed more likely to be victims than older wives.[60] Rates of abuse progressively decline with the increasing age of the woman. It is possible, of course, that more violence is directed at young women because they are more likely to be mated with young men, who are known to be more aggressive in general. Although this explanation has some merit, three sources of evidence reveal that it cannot be the entire story. First, young women married to much older men — five to twenty-five years older — are especially vulnerable to abuse, suggesting that the age of the woman is at least partly influencing men to exert coercive control. Second, young married women are more intensely mate guarded than older married women, above and beyond what results from the age of the husband and the length of the relationship. All forms of mate guarding are directed toward younger women, not just the cost-inflicting kind. Third, diverse forms of coercive control of young women occur across cultures. These include veiling, chastity belts, foot binding, claustration, guarding by eunuchs, genital mutilation, threats of violence, and assault. These practices vary enormously in their expression but almost exclusively target young reproductive-age women. Across cultures, prepubescent and postmenopausal women are less often victims of these coercive practices.[61]

WOMEN'S DEFENSES AGAINST ABUSE

From a coevolutionary perspective, victims of aggression are unlikely to be passive recipients of violence. Given the costs to victims and the prevalence of IPV, it would defy logic if women had not developed defenses to prevent becoming victims and to ameliorate or escape the harms that abusive partners inflict. Women's kin typically form important defenses. Most men will be deterred from resorting to battering if their partner has a formidable brother, father, mother, or sister in close proximity. Among the Yanomamo of Brazil, a culture in which spousal violence is widely accepted, women "dread the possibility of being married to men in distant villages, because they know that their brothers will not be able to protect them."[62]

Unfortunately, if the husband has formidable male kin of his own in close proximity, he feels freer to resort to physical violence to control his wife. Professor A. J. Figueredo argues that the balance of power between the two kin groups predicts whether IPV is or is not used to control women.[63] Changing the balance of power is the key motive behind men's efforts to isolate their partner and women's efforts to stay connected with their family and friends. Men who are controlling attempt to sever these ties.

A related explanation is needed for why men curtail a partner's interactions with her female friends. Like her kin, female friends are a woman's coalitional allies. They act as a countervailing force to resist male control. It is more difficult to abuse a woman when she has three female friends around her than it is when she's alone. Moreover, female friends often have male partners, male friends, and male kin who further strengthen a woman's alliances. All these people are capable of spreading information that can damage the abuser's reputation. If abusing a mate is a cue to low social value, for example, people may devalue or ostracize abusers. Female friendships, in short, shift the balance of power and deter a man intent on control. The cultivation of coalitional allies provides women with their most powerful defense.

Another key defense is direct physical aggression. Although most scientific research on IPV has focused on men as perpetrators and women as victims, the past few decades have witnessed a dramatic shift. Some women do assault their intimate partners, and some researchers argue that female perpetrators are as common as male perpetrators.[64] This research was ignored for many years by scholars partly because male-initiated IPV leads to greater physical injury, but also partly because it contradicted the narrative of "patriarchy" as the primary explanation.[65] If partner violence is solely an effort by men to attain and maintain power over women, how can we explain why women would ever assault their mates? As one leading researcher framed it, "Violence between husband and wife is far from a one way street."[66] And sometimes the assaults women initiate are as vicious as those men mete out — male victims report being punched, kicked, bitten, choked, and stabbed by their mates. Countering the argument that men's greater size and strength tip the balance in their favor, some argue that "the average man's size and strength are neutralized by guns and knives, boiling water, bricks, fireplace pokers, and baseball bats."[67]

The apparent clash of these facts with a patriarchy explanation of IPV has at least one partial solution — *women assault partners in self-defense or to deter future aggression.* This is the explanation proffered by the eminent evolutionary psychologists Margo Wilson and Martin Daly. They propose that the argument for sexual symmetry of marital violence is a myth if we consider the underlying motives.[68] The main motive for women, they argue, is self-defense or the protection of their children. Consider the most extreme form of partner violence — homicide. Women who kill their mates typically do so after months or years of being abused, when they have exhausted other means of deterring male aggression, when they feel hopelessly trapped, and when they fear for their lives or those of their children.[69]

An illustrative example comes from a Michigan woman named Francine Hughes. By 1977, she had suffered thirteen years of violent physical abuse and sexual assaults by her husband, James Hughes. On March 9, 1977, James came home drunk, beat up Francine, raped her, and then fell asleep on the bed. Francine put her three children in the car and told them to wait for her. She then went into the bedroom and poured gasoline on the bed on which James was passed out. She lit a match and set it ablaze. Francine then drove her children to the police station and confessed. Although she was charged with first-degree murder, the jury sympathized with her predicament and found her not guilty because of "temporary insanity." This true story was made into a movie, *The Burning Bed*, starring Farrah Fawcett as Francine. The movie, hailed as one of the ten best TV movies of all time, helped to draw attention to the pervasive problem of partner abuse and prompted more lenient sentences for women who killed their husbands in self-defense.

In sum, although some men use violence as one means of

controlling women, preventing them from having sex with other men and preventing them from leaving, women use defenses such as maintaining proximity to kin, cultivating male and female coalitional allies, and sometimes responding aggressively to defend against men's coercive control. Each sex deploys tactics to influence the other, to be closer to their own advantage. The fact that self-defense is a key motivation of women who use violence, however, does not entirely settle the scientific debate over the gender symmetry or asymmetry of IPV. Perhaps some women use IPV for reasons parallel to those that motivate men — to keep a partner faithful, to punish suspicions or observations of infidelity, and to deter the partner from abandonment.

Abuse does deter some from abandoning a relationship, leaving many victims trapped. But some succeed in leaving. Among those who succeed, however, some become victims of another form of sexual conflict, stalking by the ex — the focus of the next chapter.

CHAPTER 6

STALKING AND REVENGE AFTER A BREAKUP

> I broke up with him and he couldn't handle it. He felt like he owned me or controlled me and when I made decisions such as this, he would just snap. I could not date anyone because he would get so mad and he would try to fight that other guy.
>
> — Woman, age twenty-one, describing her ex-boyfriend who stalked her

FROM A LEGAL PERSPECTIVE, stalking is a repeated pattern of malicious conduct that inflicts costs on victims and instills fear. Other names for stalking include "unwanted pursuit behavior" and "obsessive relational intrusion." Examples of stalking include repeated phone calls or text messages, monitoring, exaggerated affection, unwanted gifts, intruding into the victim's social interactions, invading personal space, trespassing on property, following the victim around, showing up at the victim's work unexpectedly, gathering information covertly, and unwanted persistent pursuit. One unusual aspect of stalking is that it includes many behaviors that are typical in normal courtship, such as giving gifts, showing affection, and sending affectionate text messages. If these actions are welcomed, their target may construe them as signs of love and commitment or perhaps feel flattered by the attention. If they are not welcomed, however, they can range from creepy to terrifying.

California became the first state to enact laws against stalking in 1990.[1] The other forty-nine states followed in quick succession.

Laws prohibiting cyberstalking in the United States have been slower in the making. Only a third of states have anti-stalking laws that are broad enough to include cyberstalking. California again took the lead, amending its stalking laws in order to prosecute a fifty-year-old man, a former security guard, who stalked a woman who rejected his romantic advances. His crime appalled many. He attempted to solicit over the internet men to stalk and rape the woman who rebuffed him.[2] He pled guilty to the crime on April 28, 1999. Nonetheless, two-thirds of states still fail to render cyberstalking illegal.

Stalking is one of the few criminal offenses in which the *psychological state of the victim* is critical in defining the crime. Precisely the same pattern of repeated conduct is criminal if it would evoke fear in a "reasonable person" but perfectly legal if welcomed or if the victim feels psychologically unconcerned. As a personal example, when I was a professor at a former university, I received unwanted cards with love notes and other items such as coffee mugs with red hearts on them. These came reliably every Valentine's Day and on my birthday. Someone placed them in my university mailbox. All were anonymous. These coincided with unidentified phone calls. When I answered, nothing but silence greeted me. This pattern of unwanted contact continued for several years. Although I found the pattern a bit creepy, I assumed that it might be someone in my workplace who had a crush but was shy about revealing it. It would not pass the threshold into criminal stalking, however, because I felt no fear. It finally ended when I moved to another university; I never found out the identity of the perpetrator. Female professors who have experienced analogous patterns of events — and I know several who have — have suffered from higher levels of fear than I did.

An evolutionary perspective provides fresh insights into the underlying psychology of stalking perpetration and victimization.[3] It poses these two key questions: Is stalking aimed at solving specific challenges of attracting or retaining a partner? Does stalking actually succeed, at least some of the time, in solving these problems by hijacking its victim's psychology? Answering these questions requires addressing the issue of sex differences.

ARE THERE SEX DIFFERENCES IN STALKING?

It should come as no surprise that men are more likely than women to engage in more overt and aggressive forms of stalking. Women are victims more frequently than men. Stalking that rises to an illegal level in most states is perpetrated by roughly 80 percent men and only 20 percent women.[4] Estimates put the percentage who will ever become stalking victims at some point in their lives at 8–32 percent for women, but only 2–13 percent for men.[5] Using the most conservative estimates, more than a million women are stalked in the United States each year. Due to underreporting and the fact that many people do not attach the label of "stalking" to the relevant class of behaviors, the actual number of victims is undoubtedly considerably higher.

One explanation for this stark sex difference resides in the legal definition of stalking. All patterns of conduct must meet the legal criterion of *repeated behavior that instills fear in the victims*. Since women typically experience greater fear than men in response to the same acts that add up to stalking, perhaps because of men's more formidable size and strength, women's evolved mating psychology, or both, more women feel victimized and more men cross the legal threshold to qualify as criminal stalkers. Stalked men are less likely to experience fear and less likely to report their victimization to the police even if they feel frightened.

Because of the known sex differences, this chapter focuses primarily on male perpetrators and female victims, although I also touch on female perpetrators and male victims.

THE PHYSICAL AND EMOTIONAL TOLL ON STALKING VICTIMS

Stalking typically inflicts an enormous toll on victims. One study found that between 25 and 46 percent of victims experienced one or more of the following harms: anxiety, anger, irritation, stress, and sleeping disorders.[6] A study of 107 Italian nurses who had been stalked found that 51 percent suffered from anxiety, 50 percent experienced fear, 48 percent felt angry, 29 percent experienced sleep disruption, 24 percent suffered appetite or weight problems, 19 percent suffered headaches, and 19 percent experienced panic attacks.[7] Stalking victims sometimes suffer financial losses and are forced to quit their jobs. Student victims see their grades plummet; a few drop out of school. Some victims are forced to change their names, change their appearance, withdraw from social activities, or even move to a new city. Many stalkers issue threats of physical or sexual assault. Victims worry that stalkers will follow through on threats.

Threats combined with psychological uncertainty over if or when the threats will be carried out prove to be especially distressing. One woman in our study of 2,431 stalking victims, for example, reported that her stalker sent her a photo in which he had "pasted a plastic knife with red nail polish to [the picture of] my head." A study of 220 female undergraduates who had been stalked by a former romantic partner found that 62 percent of the perpetrators issued threats of violence, and 36 percent of the victims experienced actual physical violence by the stalker.[8] Other studies find that roughly 25–40 percent of stalking victims suffer some form of violence from a rejected stalker.[9] Violence comes in many forms, including slapping, punching, and kicking. Some assaults escalate to knives or guns. The two strongest predictors of a stalker actually carrying out violence are *issuing threats of violence* and a high level of *jealousy* during a relationship prior to a breakup.[10] Alcohol consumption and illegal drug use are also predictors of stalking violence.

Stalking occasionally turns deadly. One study of women killed by former partners found that 76 percent of them had been stalked by their ex during the one-year period preceding their murder.[11] There is evidence that O.J. Simpson stalked his ex-wife, Nicole Brown Simpson, prior to her murder on June 12, 1994. O.J. was extraordinarily jealous during their relationship. Nicole called the police several times for protection from his threats and physical abuse, and police have photos of Nicole's battered face showing that O.J. followed through on his threats. During one of his stalking episodes after their separation, he apparently spied her through a window having sex with another man in her Brentwood home. This may have pushed him over the edge, escalating his stalking to the point of planning her murder. (O.J. Simpson was acquitted of murder during the criminal trial but found liable for Nicole's death in the civil trial.)

The fear of assault that victims experience, in short, is entirely warranted. Women should take threats of violence from stalkers seriously since threats are often carried out. Even if stalkers do not follow through on threats, the psychological toll on victims is steep according to stalking experts: "Though perhaps counter to expectations, it appears that the sense of looming vulnerability that accompanies threats may be more productive of psychological distress in stalking victims than the reality of actual physical assault, which, importantly, may precipitate a more sympathetic response, particularly from law enforcement."[12] In short, stalking produces psychological terror.

Stalking also interferes with victims' lives and lifestyles. Victims often feel forced to relocate, change jobs, alter travel routes, and curtail socializing. Some stalkers ruin the finances of their victims, such as racking up a large debt on the victim's credit cards. Some spread malicious gossip about the victim, destroying their social reputation. Some harass the victim's children. A few harm pets. Stalking, in sum, can inflict massive damage to all aspects of a victim's life.

WHY DO REJECTED LOVERS STALK THOSE WHO SPURN THEM?

Traditional theories of stalking parallel those of intimate partner violence. The most common invoke an insecure attachment style or a psychological disorder. According to attachment theorists, those with an insecure attachment style distrust intimates, fear rejection, and show a heavy emotional dependence on a partner. They have poor social skills and become overly clingy with romantic partners. Ironically, their fear of rejection is entirely warranted. People tend to push away super-needy or overly clingy partners. They inflict a *heavy relationship load* on their partners.[13] Consequently, they bring about precisely the romantic rejection they most dread and become even more emotionally unstable upon the breakup.[14]

The attachment theory of stalking has some scientific support. A study of 2,783 college students assessed insecure attachment with these self-report items: "I'm afraid that I will lose my partner's love"; "I often wish that my partner's feelings for me were as strong as my feelings for him or her"; "Sometimes romantic partners change their feelings about me for no apparent reason"; "I'm afraid that once a romantic partner gets to know me, he or she won't like who I really am."[15] The students also completed a standard measure of stalking that included eleven items, such as following, watching, or spying on someone; standing outside their home; showing up uninvited; and sending unsolicited messages or emails. These behaviors had to be repeated with the same victim to qualify as stalking and had to be viewed by the victim as harassing, frightening, intrusive, and threatening.

The study confirmed that those who scored high on insecure attachment do indeed have a stronger proclivity to stalk lovers who have spurned them. The same study also found that self-reported stalkers also had a slightly more prevalent history than non-stalkers of psychiatric disorders, such as anxiety and depression. These links were not strong, however, leading the study's authors to question pathology theories of stalking. Nonetheless, people convicted of criminal stalking, as opposed to average college students, are indeed more likely to suffer from a psychological disorder. One study found that criminal stalkers tend to be high on the autism spectrum disorder (ASD) scale — a syndrome marked by deficits in social mind reading.[16] Those high on the ASD scale have trouble making eye contact when interacting with others, struggle with the back-and-forth of normal conversation, and have difficulty understanding another's point of view. Some stalkers have a diagnosable borderline personality disorder or narcissistic personality disorder.[17]

An interesting disconnect is that many stalkers do not construe their actions as stalking, harassing, or intrusive. Some stalkers view their actions simply as expressing genuine romantic interest and mistakenly believe that their actions are welcomed. Some even misinterpret their victims' attempts to push them away as tests of their commitment. They see persistence as a way to pass the tests, to show the strength of their love. Many are notoriously oblivious to the psychological suffering of their victims — another deficit in social mind reading.

The motives of stalkers vary widely. Some stalk to inflict revenge on a boss or work colleague who they feel has thwarted them. Some stalk their enemies as preparation for an attack. Some are sexual predators, a topic we will examine in the next chapter. The most common motives of stalkers are closely linked to mating. And among mating motives, stalking an ex when a breakup is impending or in the aftermath of a breakup is the most frequent, and is the primary focus of this chapter. These stalkers fall under the category of "The Rejected Stalker."[18]

Stalking often starts when the possibility of a breakup is looming on the horizon of a relationship. Stalking behaviors, in principle, can be effective mate-guarding tactics because they monopolize the time of a partner, as illustrated by this female victim: "He monitors everything that I do. It's like everything I do I have to think, 'Is it going to be OK?' and 'What [will happen] if I do this?' 'If I do this how is he going to react?' It's just [that] he's constantly there."[19] Intentionally monopolizing a mate's time is common among dating and newlywed couples. It reduces exposure to potential rivals, reduces opportunities for infidelity, and sends a signal to would-be mate poachers that the partner is "taken."

In support of the mate-guarding function of stalking, one study found that jealousy, envy, and distrust of the partner were identified as motives in 32 percent of stalking cases.[20] In other research, 15 percent of male stalkers were motivated to check up on their partner, trying to catch a lover consorting with an interloper. Across studies, an average of 21.71 percent of stalking included a jealousy motive, the emotion most central to mate guarding.

Although stalking a former lover can be dyadic, there are often three individuals involved — the former mate, the spurned lover, and the new mate or potential mate. The jilted individual faces two major problems — attempting a reconciliation with the former mate and thwarting the new or potential mate. Sometimes stalking succeeds. Often it fails, as illustrated by the following case.

The case involved a twenty-five-year-old man who stalked his ex-girlfriend, the first true love of his life. He reported having an intense emotional connection with her.[21] He had been inexperienced in love before that, never having had a relationship that lasted more than two weeks. He attributed his inexperience to being too shy. He became totally smitten, stating that he and his girlfriend were "soul mates" and "seemed to be made for each other."[22] After he began dating her, he quit his job. Why? In order to spend more time with her.[23] Because most women value financial stability in a potential mate, it is not surprising that this act did not exactly endear him to her. Moreover, escalating the time he wanted to spend with her signaled "high relationship load" in neon lights. Talk about a recipe for "How to Lose a Woman in Two Easy Steps"! His girlfriend, of course, ended the relationship. Her former lover oscillated between thoughts of suicide and fantasizing about getting back together with her. During the month after being rejected, he attempted to contact his ex-girlfriend dozens of times, showing up unwanted and unexpected at her workplace and home. He spied on her through her windows, hiding in the shadows outside her home. During one of these stalking vigils, he witnessed his ex-girlfriend having sexual intercourse with another man. He reported feeling "destroyed...mad and furious beyond anything I could describe."[24] He burst through the door and stabbed his rival with a knife.

This case is typical in several respects. First, getting romantically rejected by a lover is a key trigger for stalking. Second, most stalkers are men, and this is especially true of spurned stalkers, of whom 90 percent are men.[25] Third, stalking typically starts when the partner either ends the relationship or attempts to end it. Fourth, one key goal of stalking is to reunite with the partner. When a man perceives that reunion is irrevocably off the table, however, rejected stalkers sometimes seek revenge — a topic we examine at the end of this chapter. Fifth, rejected stalkers tend to be the most dogged and intrusive of all stalkers. They sometimes stalk for months or years following a romantic breakup.

Rejected stalkers experience two key emotions — intense *rage* and a profound sense of *humiliation*. Rage, as we saw in previous chapters, often has the goal of recalibrating the target's welfare trade-off ratio, and sometimes it works. Victims of enraged stalkers, fearful of violence, accord them with power and sometimes accede to their wishes. Humiliation and shame also suffuse the rejected stalker's phenomenology. Spurned lovers suffer a decline in perceived mate value since people suspect that there must be something wrong with the one who was rejected. In addition, former mates may share status-damaging secrets about their exes with friends and potential romantic partners, such as revealing their insecurities, personality problems, or sexual deficiencies, decreasing the spurned partner's ability to attract future mates.

Shame and humiliation are evolved emotions that track our social reputation, the status in which we are held by others in our group.[26] Feeling humiliated is psychologically painful but supremely functional. It motivates attempts to repair reputational damage. It spurs efforts to avoid status-harming situations in the future. And it activates efforts to reunite with a lost loved one in order to stanch the loss of status.

The case I described earlier captures several other key qualities of rejected stalkers. Their victims tend overwhelmingly to be young women. Most studies find that women of peak reproductive age — between eighteen and thirty — are most likely to become victims of stalking. Research that Professor Joshua Duntley and I carried out on 2,431 victims of stalking confirmed this finding with a disturbing twist. We found that females are much more likely to become victims of stalking after they go through puberty.[27] Corroborating our findings, a study of 18,013 high school students in the state of Kentucky found that 19 percent of women reported being victims of stalking within the past year, beginning during adolescence.[28] The rates were roughly comparable across the ninth, tenth, eleventh, and twelfth grades, suggesting that many women fall victim to stalkers when they are in their teens. As with partner violence, male stalkers are typically older than their female victims.

Young women just entering the mating market sometimes fail to accurately perceive their mating desirability. Consequently, they are more vulnerable to romantic approaches by men of lower mate value — often older men who exploit the youth's inexperience and naïveté. With increased life experience and maturity, young women come to more accurately appraise their desirability on the mating market, motivating them to reject a partner when they perceive they can trade up. Jilted men who stalk former lovers perceive, often accurately, that their lost partner is *irreplaceable.* Irreplaceability comes from the fact that stalkers are less desirable on the mating market than their victims. In our study of stalking, victims overwhelmingly report that their stalker would have more difficulty finding someone to date than they would. They also report that their stalker is considerably less attractive to the opposite sex than they are. Stalking victims unquestionably believe that the lovers they spurn are beneath them in mating desirability, or at least they come to this realization over time.

One victim in our study stated, "I think he was just scared that he wouldn't find anyone else to love him." Another observed, "He was so lonely that when I showed him a little bit of attention, he just got hooked on it and made me the focus of his life." A third said, "He didn't have any friends and was laughed at by the whole school. He usually tried to get dates with girls that were out of his league." So the stalker, being lower in mate value than his ex-partner, realistically perceives that it will be difficult or impossible to replace her with a mate of comparable value. "Since she loved me once," he thinks, "perhaps I can win her back."

Stalkers implement desperate measures to reunite with their lost lovers. Some become suicidal. One victim in our study reported that "he threw himself in front of a car after we broke up the first time." This stalker succeeded in luring her back, at least temporarily. Some issue threats. One stalking victim said, "He told me that if he couldn't have me, no one else could," a sentiment that signals great danger, so women should take these declarations seriously. Another threatened violence to her pet: "He told me that if I broke up with him, he would cut off my dog's head." Sometimes a threat can be made with just a glance that might be unnoticeable to outsiders. One woman reported that a certain look or facial expression from her ex could evoke fear: "Seeing... my ex-husband brings back memories.... He has a way of looking at me, he knows how to look at me when he sees me to make me just shudder. And I'm afraid."[29]

Stalkers sometimes appear harmless to outside observers: "A man stands quietly leaning against his car in a parking lot across from an office building. He smiles and greets passersby and appears innocuous. No one really notices him—except his ex-wife who works in the building across the street. She notices that he appears here every day just watching the building and watching her coming and going and, with his unwanted calling and his history of assaulting her, his presence is both a warning and a threat. She worries constantly about her safety. He is stalking her, yet what he is doing to her is invisible to the public, the police, and the courts."[30]

Rejected stalkers often aim their threats at potential mates of their ex: "It was really hard for me to start dating other people for fear that either I was going to be hurt by the stalker *or the person I was dating would be hurt by the stalker.*" Another said: "I only tried to get involved with one other dude while this was going on. My ex contacted him and told him that I was already taken and threatened him. This scared him away." Stalkers sometimes succeed in frightening potential mates away, leaving victims feeling like they are imprisoned by invisible bars: "This person wouldn't leave me alone, and the overall situation discouraged me from trying to date anyone else. I felt like I was trapped." Many potential mates are deterred from courting a woman who has a stalker who won't let go, fulfilling the stalker's goal of driving away rivals. Just as perpetrators of intimate partner violence try to cut off their victim's social ties, stalkers try to scare away potential mates who might otherwise serve as bodyguards. Stalking sends a strong signal to rivals to "stay away" and so functions as a form of mate guarding.

It may strain credulity to posit that stalking has actual adaptive functions in an evolutionary sense. After all, it frequently fails to achieve its intended goals. In the modern environment, it can land the stalker in jail. Stalkers suffer damage to their status and reputation, compounding the costs. The stalking tactic may or may not succeed in its intended outcome in any given instance. But if it sometimes succeeds, or more accurately if it has sometimes succeeded over the millions of instances in which it was attempted across evolutionary time, then selection could favor an adaptation to stalk former mates because it resulted in an on-average reproductive gain compared to other courses of action.

Our study of 2,431 stalking victims found that stalking does sometimes succeed, even if the victory is a temporary one. A full 30 percent of women victims agree to meet with their stalkers at their request; 13 percent agree to date them; and 6 percent agree to have sex with them. These numbers are quite conservative in that our study was biased toward including victims of unsuccessful stalkers, so the actual success rates are likely to be higher. Even considering these percentages, however, a 6 percent success rate in regaining sexual access may be better from the stalker's perspective than the alternative future of becoming involuntarily consigned to total celibacy. Because the stalkers in our study are lower in mate value than their victims, even low success rates can render terror-inflicting tactics worthwhile from the stalker's perspective, even if it comes at a large cost to the victim from her perspective.

HOW STALKING WORKS

Several factors contribute to the effectiveness of stalking behaviors in instances where it works to the stalker's advantage.[31] First, stalkers' repetitive pattern of behavior has the effect of consuming substantial portions of their victims' time — time that could be devoted to solving other problems. This loss of time can represent a significant opportunity cost to the victims of stalking, making them less likely to be successful in other domains of their lives, such as attracting mates, retaining current mates, managing other social relationships, and getting ahead at work. In our study, victims of stalking suffered all of these costs — disruption of work, education, and romantic life.

Second, stalking consumes the victims' psychological resources, requiring that their attention be devoted to their stalkers and their conflicts with them. This creates cognitive opportunity costs, hijacking victims' psychological space. It shunts attention and effort away from the psychological effort needed to think through other tasks relevant to their own life goals.

Third, because stalking decreases the amount of time and thought victims can devote to other social relationships, it can have the effect of socially isolating them. Women are especially vulnerable to this cost. Our study revealed that whereas 37 percent of male victims reported that their stalker disrupted their social life, an astonishing 70 percent of female victims reported experiencing this harm. Isolation combined with driving away romantic alternatives may lead victims to perceive that a relationship with their stalker may be better than no relationship at all.

Stalking also sets up what psychologists call a *negative reinforcement contingency*. At an abstract level, there are two ways to reward someone's behavior — providing abundant benefits or stopping the infliction of a barrage of costs. The stalker causes harms in the form of repeated phone calls, texting, and menacing lurking. When the victim gives in to the stalker's demands, the stalker curtails these repeated costs, thereby rewarding the victim when she complies with his entreaties. Unfortunately, submitting to the stalker's demands, in turn, rewards him for using the cost-inflicting strategy. Consequently, stalking returns quickly when the victim no longer complies. In short, some victims give in to their stalkers' demands to avoid the catastrophic harms they can wreak.

REVENGE PORN

When victims fail to comply, some stalkers turn to *revenge porn* — posting sexually explicit images or videos of the former partner on the internet without their permission. Unfortunately, revenge porn sometimes works. Some stalkers use the threat of revenge porn to coerce their exes into having sex. Some use the threat of revenge porn to force victims into resuming their relationship in order to prevent the stalker from posting the sexually explicit images. When that fails, the stalker's goal is simply to inflict as much damage on the victim as possible.

One real-world case exemplifies revenge porn. Jane (a pseudonym) consented to her boyfriend's request to photograph her in the nude.[32] He assured her the pictures were solely for his own viewing pleasure. After their breakup, he violated his promise and posted one nude image of Jane on a revenge-porn internet site and even added her actual contact information. Jane then got barraged with emails, Facebook friend requests, and phone calls from strangers, mostly from men seeking sex. When she reported these events to the police, the responding officer told her that there was nothing that could be done. The boyfriend had photographed her with her consent. He had only posted a single nude photo. Stalking and harassment laws typically require repeated instances to qualify as criminal. This occurred in the year 2013, prior to laws against revenge porn.

Revenge porn can devastate women's lives. One study of eighteen female victims found that these online betrayals evoked severe anxiety, depression, thoughts of suicide, and PTSD.[33] Victims sometimes lose their jobs when bosses or co-workers discover the nude images, circulate them, and spread gossip about the victim. Companies often refuse to hire victims of revenge porn. Women live in fear that their friends or family will come across the images. Their social reputation suffers.

Rebekah Wells founded the organization Women Against Revenge Porn after she became a victim. After a breakup, her ex-boyfriend posted photos. She discovered "an online gallery of nude photos of myself after Googling my name." She complained to the police. The handling officer assured her after several months that he would protect her, that her case had been closed, and that the photos presumably had been removed. Then he asked her out on a date. After a brief relationship, Wells broke it off, at which point the policeman threatened to post the sexually explicit images of her that he had obtained from her first complaint. She eventually received justice of a sort when the cop was subsequently fired. But she suffered a couple of years of anxiety, depression, severe weight loss, and a host of other psychological symptoms.[34]

Tragically, revenge porn can trigger stalking by other men. This fact was revealed on one revenge-porn site that allows women to comment on the images. One woman pleaded with the webmaster: "I am being stalked at work. PLEASE, PLEASE, PLEASE remove my photos." Because the images can be downloaded easily by the thousands of visitors, even removed images quickly reappear on other revenge-porn sites, which prompts other men to stalk.

The psychological devastation women experience stems in part from damage to a woman's *sexual reputation*. This reputation, in turn, is formed and driven by two key evolutionary forces of sexual selection — men's mate preferences and women's intrasexual competition. Men worldwide prioritize sexual fidelity in long-term mates. Men interpret cues to perceived promiscuity as compromising prospects for fidelity in a committed partner. In contrast, men are attracted to cues of a woman's perceived promiscuity when they seek casual sex partners because these cues convey information about their chances of succeeding sexually. So victims of revenge porn suffer damage to their long-term mate value in the eyes of men. Women perceived as promiscuous, even if that perception is entirely erroneous and based on images they themselves have not posted, also tend to be slotted in the male brain as potential short-term mates.

Victims suffer from other women as well. When women compete with one another for desirable mates, they do so in part by striving to embody what men want more than their rivals do. Women compete with one another for a desirable sexual reputation, which leads them to derogate other women, sometimes viciously, by slandering their rivals with labels such as "slut," "slag," "whore," and "tramp." Women also ostracize women perceived as promiscuous because they do not want their own reputations tarnished by association. Victims of revenge porn suffer the indignity of lower perceived mate value and an onslaught of men who stalk them for casual sex.

In short, revenge porn hijacks women's psychology by damaging their sexual reputation, their perceived mate value, their future long-term mating prospects, and their alliances with other women. It even threatens their family ties, especially in conservative cultures in which women's chastity is prized above all other perceived virtues.

There are some signs of hope of deterring men from posting sexually graphic pictures of their exes. As of 2019, forty-six states plus Washington, DC, have enacted laws against revenge porn.[35] New York State, the forty-sixth to enact such laws, imposes penalties of up to one year in jail and a $1,000 fine. And since roughly 80 percent of nude images are actually selfies women take and share with their boyfriends, the women own the copyright to the images. Revenge porn, therefore, often violates copyright law, and perpetrators now can be sued in civil court.

Because revenge porn has only recently become a crime, few studies have been conducted to determine how often it occurs, who perpetrates it, who are its victims, whether the laws are effective deterrents, and the magnitude of its shattering aftermath. In a rare exception, a study of 4,122 Australians found that roughly 10 percent had been victims of some type of revenge porn, someone either threatening to post nude photos online or actually doing so.[36]

It will surprise no one that men were twice as likely as women to admit to perpetrating it. Young women, in the age range of eighteen to twenty-four, were most likely to be victimized — similar in age to victims of partner abuse and stalking. Australian victims reported feeling angry (64 percent), humiliated (55 percent), suffering a loss of self-esteem (42 percent), experiencing depression (40 percent), and being afraid for their safety (32 percent). Damilya Jossipalenya became the first known victim of suicide as a result of revenge porn. Her boyfriend sent a video of her performing a sex act on him in a telephone booth to one of his friends and then threatened to send it to her conservative family back home in Kazakhstan. Included in his threat were the words: "I warned you not to fuck with me."[37] She jumped from a third-story balcony to end her suffering and her life.

The horrors victims experience explain why Rebekah Wells argues that revenge porn might be more aptly labeled as "cyberrape."

DEFENSES TO COMBAT AND END STALKING

Victims of stalking have a number of defenses at their disposal. Professor Duntley and I developed a website specifically to help victims — www.stalkinghelp.org. It is available in both English and Spanish, so if you know someone who is a victim, you can refer them to this website for detailed information about victim services, legal remedies such as restraining orders, and links to other helpful victim websites. Another excellent resource is the work of Michele Pathé, whose book *Surviving Stalking* remains one of the best sources of advice for stalking victims and those who care about them.

This section highlights a few defenses that may help, with the important qualification that no defense works 100 percent of the time, and some stalkers are notoriously persistent in circumventing these defenses. The fact that the typical stalker perpetrates their crime for roughly ten months, and some persist for years, defying restraining orders and the threat of jail time, is a testament to their relentlessness. Rejected romantic stalkers, as noted earlier, are among the most persistent. Nonetheless, knowledge can help empower stalking victims.

Enlisting social support is one of the first and most important lines of defense and perhaps the most evolutionarily ancient. Friends and family members can function as bodyguards. Humans are an intensely social species. We evolved in small-group living, surrounded by kin and allies. The amount of social isolation in the modern world has created a profound mismatch, but it can be remedied.

Gathering social support deters stalkers in several ways. It sends a signal that the victim is not alone. Social allies can also bear witness, providing independent corroboration of the stalker's actions. They can call the police should the stalking become dangerous. They can become witnesses should legal proceedings take place. And they can rally their own social allies should more reinforcements be needed. Let close friends or family know your schedule and where you will be. Some stalking victims feel too ashamed to alert allies; some blame themselves for the stalker's conduct. Overcome these barriers and activate your social network.

Ceasing all contact with the stalker is one of the strategies most frequently recommended by stalking experts.[38] Stalkers often construe any interaction as rewarding, even if it's negative. Reasoning and logic rarely work. They give the stalker hope that the romantic relationship can be renewed. If the stalker calls fifty times, do not answer; have the number blocked. Pleading with the stalker to stop rewards him with contact. Angry outbursts at the stalker, ironically, also can inadvertently give him the attention he seeks. Do not return gifts; the stalker will view this as engagement. It must be recognized, of course, that cutting all contact can be difficult for stalking victims to implement since they worry that doing so might enrage the stalker.

Equip yourself with defenses. Have emergency numbers on speed dial. Make sure your apartment or house is secure, doors deadbolted and windows locked. Change your locks if the stalker has a key or can access one. Use shades or drapes to block visual access from the outside. Install a peephole on your door to verify the identity of anyone who knocks or rings. Consider installing sensor lights and a motion-activated video camera to monitor unauthorized activity outside your domicile. Hire a home-security service if you can afford it. Some stalkers slit tires or throw acid on cars. Keep your car locked and in a well-lit area. Equip it with a loud alarm should the stalker try to break in or tamper with it. Consider taking a self-defense class. Carry Mace if it is legal in your area, and have it ready to use if you find yourself alone and unguarded.

Document the stalker's conduct. Stalking is a crime that requires a pattern of repeated conduct to qualify as criminal. Keep a running diary documenting every instance of that conduct, whether on your cell phone, on a computer, or in a handwritten journal. Keep letters, text messages, emails, and gifts from the stalker, documenting when they arrived, establishing a chronology of the criminal conduct. Do not destroy the evidence; it will be needed should it come to legal proceedings.

Minimize your exposure through social media. Many stalkers cyberstalk. They can monitor your activities when you post on Facebook, Twitter, Instagram, or other social media. They can use information to track your whereabouts, your travels, your interests, and your activities. They can learn who your friends are and exploit that information to infiltrate your life or to spread malicious gossip about you. Consider staying off social media as much as possible.

Notify the police. Take stalking seriously. It can be dangerous, especially from rejected former intimates. Seek advice. Police have become more sensitive to crimes of stalking and many are now experienced in dealing with stalkers. Provide police with the evidence and documentation you have assembled, including the name, address, contact information, and photos of the stalker. Provide information about the stalker's car — make, model, color, and license plate if known. Police cannot act without evidence. Consider obtaining a protective restraining order if warranted. They deter some stalkers. And although persistent stalkers sometimes violate these orders, those violations provide grounds for more serious legal sanctions. In some severe cases, it takes imprisonment to stop the stalker.

Change your name, address, and identity if it means saving your life. Although only a tiny minority of stalkers murder their victims, a few do. The police rarely have the time or staff to protect a victim from a persistent stalker hell-bent on revenge. If your life is in danger, drastic defenses may be needed. Most stalkers eventually move on.

Although some women suffer partner violence during a relationship and stalking in the aftermath of a breakup, many more women experience forms of sexual coercion. Coercion ranges from sexual harassment in the workplace to nonconsensual sex at the harsh hands of strangers, acquaintances, friends, and romantic partners — topics to which we now turn.

CHAPTER 7

SEXUAL COERCION

> Rape is one of the most terrible crimes on earth and it happens every few minutes.... What really needs to be done is teaching men not to rape. Go to the source and start there.
>
> — Kurt Cobain

SEXUAL COERCION IS sexual activity that ranges from unwanted sexual attention through more severe forms of sexual harassment and rape via pressure, incapacitation, threat, or force. I want to pause here to underscore that I am not on a quest for a competitively "superior" answer to the question "Why does sexual coercion happen?" I am not suggesting that we need only attend to evolutionary angles on the phenomena — because so many valuable studies have shed important light on the cultural, social, and environmental circumstances that contribute to patterns of sexual coercion. Instead, I suggest that it is important to supplement such perspectives with all the best tools at our disposal, and one of them is to examine sexual coercion from an evolutionary perspective.

CAN AN EVOLUTIONARY PERSPECTIVE HELP?

Because the topic of sexual coercion understandably raises a wide variety of emotions, especially given the wide range of reader life experiences, it is crucially important to begin with a few caveats to help keep the discussion on a track that is simultaneously humane and constructive. This is especially true in the context of evolutionary perspectives on sexually coercive behavior, because there have been some vivid and high-conflict reactions to this topic in the past.

First, an evolutionary perspective does *not* view sexual coercion as a biological imperative, as inevitable, or as ineluctable. Just as modern science has created novel vaccines and drugs to eliminate many "natural" diseases, with enough knowledge we can create personal, social, and legal environments that curtail or suppress the components of male psychology that contribute to sexual coercion.

Second, to say that there may be evolutionary influences on the probability of certain behaviors never says anything, by itself, about whether those behaviors are good or bad from a moral perspective. That is especially true in the context of sexually coercive behaviors that are justly condemned for good reasons. In the same way that studying the multiple causes of cancer does not mean that one thinks cancer is a good thing, studying the multiple factors contributing to sexual coercion doesn't mean that sexual coercion is in any way acceptable.[1] Identifying evolutionary origins of nefarious behavior in no way justifies or excuses it.

The legal scholar Owen Jones notes that generally speaking, laws, in this case laws against sex crimes, are levers designed to influence human behavior.[2] They are designed to deter people from doing things we don't want them to do (e.g., rob, murder, rape) and to encourage people to do things we want them to do, such as reporting observed criminal activity and providing honest under-oath testimony to those crimes. For laws to be maximally effective, Jones argues, their designers must have an accurate model of human nature. The more accurate the model, the more effective the legal levers. An evolutionary perspective clearly does not solve all problems, of course, and the science is far away from a complete and accurate model of human nature. An evolutionary scientific lens, however, sheds light on human nature, on the nature of male and female sexuality, and on the psychology of sexual coercers and their victims. It offers the potential for more effective policies for reducing sexual misconduct ranging from unwanted sexual attention to sexual assault, and my hope is that it will be used for this purpose.

UNWANTED SEXUAL ATTENTION

Attention is a sharply limited resource. From among the thousands of things we could focus on — a twig on a tree, an ant crawling into a crack, one of a million grains of sand on the beach — our evolved psychology ignores most of them. Focusing on one thing necessarily precludes focusing on other things. Evolution has sculpted our brains to devote this limited resource to stimuli maximally relevant to solving pressing adaptive challenges. When we are hungry, the sights and smells of fresh-cooked pizza or a tray of brownies monopolize our senses. In the mating domain, when romantic or sexual motives are activated, desirable potential partners compel our attention.

Psychologist Jon Maner and his colleagues conducted studies on *attentional adhesion* — the degree to which different visual stimuli capture and maintain focus.[3] Participants in the studies were first asked to write about a time in their lives when they were sexually and romantically aroused — primes designed to activate mating adaptations. Different images then were presented in the center of the computer screen — an attractive woman (as pre-rated by a panel of people), a woman of average attractiveness, an attractive man, or a man of average attractiveness. Following this exposure, a circle or a square popped up randomly in one of the four quadrants of the screen. Participants were instructed to shift their gaze away from the central image as soon as the shape appeared elsewhere on the screen and then to categorize it as quickly as possible as being either a circle or a square.

Men exposed to the image of the attractive woman had difficulty detaching. They took longer to shift their gaze away and longer to categorize the circles and squares correctly. Their attention adhered to the attractive woman. Some men, however, succumbed to attentional adhesion more than others. Men inclined to pursue a short-term mating strategy got especially stuck. These were men who tended to agree with statements like: "I can imagine myself being comfortable and enjoying 'casual' sex with different partners" and "Sex without love is OK."[4] Heightened attentional adhesion to attractive women stands out as a special feature of the psychology of men seeking casual sex. It appears to serve a specific function — identifying women who might be potential sex partners. This research provides the first clue that men, especially those interested in casual sex, have an attentional bias that contributes to unwanted sexual attention.

In a related study exploiting the technology of functional magnetic resonance imaging (fMRI), scientists sought to identify the reward value of, or pleasure experienced by, viewing different images.[5] They exposed heterosexual male participants to four sets of faces differing in attractiveness, as determined by prior ratings: attractive females, average females, attractive males, and average males. While participants viewed these images, their brains were neuroimaged in six regions. When men looked at attractive female faces, the *nucleus accumbens* area of the brain became especially activated. The nucleus accumbens is known to be a fundamental region of reward circuitry, a pleasure center in the brain. This reward circuit was silent when men looked at either average female faces or any of the male faces. Beautiful female faces, in short, are especially rewarding to men, psychologically and neurologically. Just as their attention adheres to attractive women, their brains reward them for looking.

Unwanted attention might not seem distressing to all people, but the fact that it can herald unwanted sexual action makes it much more menacing. Researchers of 1,005 individuals attending electronic dance music parties in New York City asked whether they had experienced nonconsensual touching, groping, or kissing at these events. Attendees at these kinds of events often consume alcohol and the drug ecstasy.[6] Although only 15.2 percent reported receiving unwanted sexual contact, women were twice as likely as men to become victims. Among women, those age eighteen to twenty-four were nearly three times as likely to be touched or groped as those age twenty-five to forty. Women victims reported that 99.5 percent of the unwanted contact was all or mostly at the hands of men. Studies of victimization at other dance venues and music festivals reveal similar patterns. One from the United Kingdom found that 27 percent of women reported unwanted forceful dancing such as grinding and rubbing, and 17 percent reported nonconsensual touching of their breasts, buttocks, or genitals.[7] Younger women, age eighteen to twenty-four, reported higher rates. Again, most perpetrators were men.

A study from Sweden revealed similar results, with 24 percent reporting unwanted sexual contact at nightlife venues.[8] Women experiencing such contact typically feel embarrassed, angry, and violated. Another survey of fifty thousand Swedes found that 42 percent of women (compared to 9 percent of men) report that they have experienced sexual harassment.[9] This jumps to 57 percent for Swedish women age sixteen to twenty-nine. Swedish women also experience substantial rates of sexual harassment while serving in the military (ranging from 9 percent to 64 percent depending on the wording of the questions) and while in high school, with 49 percent of women reporting that it is a problem.[10] Men's sexual attention to women, of course, does not always lead to unwanted sexual contact. Most men are perfectly capable of keeping their lustful impulses in check. But enough men act nonconsensually on these sexual desires to inflict harms on many women. Residing in sexually egalitarian cultures such as Sweden apparently does not eliminate unwanted sexual contact from men.

Why would men ever think an uninterested woman would be sexually interested in them? One piece of the puzzle is that the *male sexual over-perception bias* proves to be especially strong with physically attractive women. Recall that many men erroneously infer sexual interest when a woman smiles at them or incidentally brushes against their arms. Some men are shy, anxious, or inhibited, which interferes with their ability to initiate courtship when they experience sexual or romantic attraction. The sexual over-perception bias helps men overcome these barriers. If a man perceives, correctly or not, that a woman is interested in him, he will be less fearful that his advances will be rejected. Assuming that women are interested elevates a man's confidence, motivating bolder approaches — but when his belief conflicts with reality, the result is sexual harassment.

Because the targets of these advances don't know how the man would react if rejected, they attempt to deflect unwanted advances with tact and politeness in an effort to avoid conflict, preserve a professional working relationship, and prevent retaliation. These tactics, including pretending to ignore sexual advances or claiming prior commitments to avoid dates, can unintentionally backfire if the man misinterprets them as potential sexual interest. Even friendliness or professional courtesy can be misinterpreted as a sign of potential interest. This toxic mix puts victims in a terrible bind. Women risk continued unwanted sexual attention on the one hand and retaliation from spurned men on the other.

Some men have difficulty believing that the sexual attraction they feel toward a woman is not reciprocated. Attentional adhesion to women, rewarded by the nucleus accumbens and combined with the sexual over-perception bias, leads to unwanted sexual attention. When it occurs in the workplace, it also can lead to sexual harassment.

SEXUAL HARASSMENT AND AN EVOLUTIONARY MISMATCH

In the hunter-gatherer lifestyle that characterized 99 percent of human evolutionary history, there existed a division of labor by sex for at least some of the work needed for survival. Ancestral men often formed coalitions for hunting large game — a risky mode of securing high-reward food. Women spent more time gathering reliable food sources like berries, nuts, and tubers and occasionally hunting small game such as rabbits or squirrels. In some traditional cultures, women's contribution often made up 60–80 percent of the total calories for the family.[11] Both sexes contributed resources — both women and men worked equally hard — but their acquisition modes carried different risks.

Men's work was often more dangerous. A swift kick from an angry game animal who did not want to be turned into a meal could injure or kill a hunter. Intergroup warfare, another risky activity, was almost exclusively a male endeavor. It required defending the group against attacks from hostile outgroups and launching offenses against those outgroups when opportunities tilted in their favor, such as when outnumbering the other group. Even nonfatal injuries, such as an arrow wound in the thigh, could turn deadly from subsequent infection. In some traditional cultures, such as the

Gebusi of Papua New Guinea and the Yanomamo of Venezuela, as many as 30 percent of men died from intergroup warfare.[12] Warfare was part of men's work.

Childcare is another form of work historically sharply divided by sex. Even in traditional cultures such as the Aka of the Republic of Congo in Africa, sometimes called the culture of "mothering men" because men interact a lot with their children, women still do the bulk of childcare. Breastfeeding, of course, is exclusively a female endeavor, so women, their female kin, and sometimes their female friends have more or less sole responsibility for newborns, infants, and toddlers. The evolutionary importance of women's role in alloparenting, building and maintaining social networks, cooperation, and peacekeeping has been increasingly recognized in recent years.[13]

Modern work environments differ dramatically from those of our ancestral past. Although a few professions are heavily dominated by one sex — such as home healthcare providers (89 percent women), social services workers (85 percent women), and brick masons, concrete workers, and auto mechanics (all more than 99 percent men) — many modern workplaces are sexually integrated.[14] Modern universities and workplaces in fields such as publishing, the news, and the movie business all contain large numbers of young women and often somewhat older men who work together. In my own field, graduate students in PhD programs in psychology typically are more than 70 percent women.[15] Combining a large pool of young women with a smaller pool of senior men, as often occurs in PhD programs and in some business sectors, produces an evolutionary mismatch, creating a perfect storm in which mating mechanisms can go awry. Evolved mating psychology illuminates why. As men get older, they are attracted to women who are increasingly younger than they are.[16] Young women occasionally reciprocate that attraction, but rarely as the age gap exceeds a decade, although of course there are exceptions.[17] So in these workplace contexts, many men are attracted to women who are not attracted to them. Most men do not act on their attractions. Most instances of sexual harassment are perpetrated by a minority of men who serially harass. Stricter workplace sexual harassment policies can help deter these men so that the burden of dealing with unwanted sexual attention does not fall unfairly on women, as it has traditionally.

Like stalking laws, sexual harassment laws typically require a repeated pattern of behavior rather than a single instance, unless the instance is particularly severe, although increasingly strict internal company policies often have tougher standards.[18] And like stalking laws, sexual harassment violations are partly defined by the psychological state of the victim — the behavior typically must be seen as unwelcome or offensive to qualify. A persistent pattern of flirting with a co-worker or complimenting them on their appearance, for example, qualifies as sexual harassment if it is unwelcome, but not if the receiver feels indifferent or flattered by the comments.

Researchers and legal scholars have identified at least two partially distinct forms of sexual harassment. The first is quid pro quo sexual harassment. Examples include offering a prized job or pay raise in return for sexual favors and threats of punishment for lack of sexual cooperation.[19] The convicted movie mogul Harvey Weinstein provides a vivid example. He allegedly demanded of one actress who auditioned for a part to show him her breasts. She declined. He replied: "Do you know who I am? You know I can make your career or I can break your career? I can make it so you will never work in this business again. So show me your breasts."[20] Quid pro quo sexual harassment is often viewed as the most severe form.

The second major type of sexual harassment includes making lewd remarks; unwanted attempts to establish a sexual or romantic relationship; unwelcome seductive behavior; unwanted touching of arms, breasts, or buttocks; and unwanted staring, leering, or ogling. In contrast to quid pro quo harassment, this cluster is marked by *sexual persistence*.[21] Although threats of cratering a career in quid pro quo harassment can be psychologically devastating to victims, sexual persistence can be no less traumatizing — the proverbial "death by a thousand cuts." Some women feel trapped in an enclosed work environment with a sexually persistent serial harasser. Unlike sex-segregated ancestral work environments in which women had female friends and kin around to deter unwanted harassment, modern women are sometimes stuck in workplaces in which their opportunities to avoid a sexual harasser are sharply constrained. Their psychological distress intensifies when they perceive the possibility that sexual harassment might escalate to sexual assault.

An evolutionary perspective offers insight into the psychology that influences sexual harassment and the circumstances that increase or decrease the probability of its occurrence. Sexual harassment is sometimes motivated by the desire for short-term sexual opportunities, sometimes motivated by a search for a lasting romantic relationship, and sometimes motivated by the desire to demonstrate or maintain power. Scholarly attempts to reduce men's bad behavior to a single motive are naïve; sex, power, and status often mingle in men's minds and cannot be neatly siloed.

Victims of sexual harassment are not random. A study of ten thousand sexual harassment complaints in the United States in 2017, for example, found that 83 percent were filed by women, in contrast to only 16.5 percent filed by men.[22] Often the male victims were harassed by other men, although there is some evidence that women perpetrate sexual harassment against men as well.[23] Among complaints filed in Canada under human rights legislation, ninety-three cases were filed by women and only two by men. In both cases filed by men, the harassers were men rather than women.[24] The fact that women are generally the victims and men the perpetrators should surprise no one, but it is an important finding that demands explanation.

Male power and patriarchy are clearly part of the picture. Men historically created the workplace rules and influenced social norms that overlooked sexual harassment. An evolutionary perspective highlights an underlying sexual psychology that influences these male-biased practices. Studies by psychologist John Bargh and his colleagues, for example, explored the unconscious links between power and sex.[25] One study found that men experienced an unconscious association between the concepts of power and sex, but this occurred only for men who scored high on a "likelihood to sexually harass" scale. In these men's minds, concepts like "authority" and "boss" were automatically linked with concepts like "foreplay," "bed," and "date." Their second study primed men to think about power and subsequently asked them to rate the attractiveness of a female confederate in the room who the men believed was just another study participant. Again, only men scoring high in likelihood to sexually harass viewed the woman as especially attractive and expressed a desire to get to know her. In short, power and sex are linked, but primarily in the minds of a subset of men. This may explain why only a minority of men in positions of power over women sexually harass them; many men with power do not.[26]

Those who do harass women sometimes erroneously infer that their attraction is reciprocated by the woman. From the woman's perspective, however, she is being friendly and deferential primarily because men in positions of power can inflict large harms or confer large benefits on their careers. Victims often interpret sexual overtures as motivated by power, since the situation holds no sexual interest for them. Men often do not view their overtures as exploitative. Harassers in positions of power often overestimate their own attractiveness and feel entitled to sexual favors from subordinates — a topic to which we return when discussing rape.

Although feminist theorists persuasively argue that some men use sexual harassment as a means of gaining or maintaining power over women or showing off their power to other men, an evolutionary perspective suggests that the reverse causal arrow may be equally true — men strive for status and power, in part, in order to get sex.[27] As the wealthy Greek tycoon Aristotle Onassis observed, "If women didn't exist, all the money in the world would have no meaning."[28]

Women's evolved sexual psychology is also critical. Women experience greater distress than do men in response to acts of sexual aggressiveness.[29] Consequently, women are more likely than men to file official complaints when harassed because they both are more harassed and experience harassment as more upsetting.

Perpetrators of sexual harassment tend to target young single women. Women over forty-five are less likely to be victims.[30] One study found that women between the ages of twenty and thirty-five filed 72 percent of the complaints of harassment, whereas they represented only 43 percent of the workforce at the time. Women over forty-five, who represented 28 percent of the workforce, filed only 5 percent of the complaints.[31] Perpetrators also target single and divorced women more than married women. In one study, single women represented only 25 percent of the workforce but filed 43 percent of the complaints; married women, comprising 55 percent of the workforce, filed only 31 percent of the complaints.[32] There may be several reasons for these relationship status differences. Spouses sometimes function as "bodyguards," deterring would-be harassers or making it more costly for them to harass. Men also may perceive single or unattached women to be more receptive to sexual advances.

Reactions to sexual harassment also follow evolutionary psychological logic. When men and women were asked how they would feel if a co-worker of the opposite sex asked them to have sex, 63 percent of the women said that they would be insulted, while only 17 percent said that they would feel flattered.[33] Men's reactions were roughly the opposite — only 15 percent would be insulted, and 67 percent would feel flattered. In some situations, where women see danger, men may be seeing sexual opportunity.[34] These reactions reflect human sexual psychology — men generally have more positive emotional reactions to the prospect of casual sex. Some women, of course, enjoy some types of attention from some men in some circumstances. Most, though, react more negatively than do men to being treated solely as sex objects.

The attractiveness of the perpetrator and the sociosexual orientation of the person evaluating acts of harassment also figure prominently. In two studies involving 1,516 individuals, participants evaluated different hypothetical scenarios occurring in a coffee bar in the workplace. A co-worker of the opposite sex enters the room and, after some small talk with the victim, abruptly makes either an implicitly sexual comment such as "When you feel tension after work, I could help you relax" or an overt sexual advance such as "fondles your back" or "grabs your butt."[35] Participants rated how disturbing each scenario would be to the targeted person and how uncomfortable the situation would be if they were in that situation. Men perceived all of these workplace sexual advances as less disturbing than women did. Both men and women high in desire for casual sex viewed the sexual advances as less disturbing. And women evaluated sexual advances from a physically attractive man as significantly less disturbing than advances from a physically unattractive man. Workplace sexual advances from men low in desirability, apparently, are more upsetting.

Women's reactions to sexual harassment also depend heavily on whether the motivation of the harasser is perceived to be sexual or romantic. Sexual bribery, attaching job promotions to sex, and other cues that the person is interested only in casual sex are more likely to be interpreted as harassment than are signals of genuine romantic interest such as compliments on hairstyle or mild flirtation.[36] College women view acts by co-workers such as a man putting his hand on a woman's genital area or trying to corner a woman when no one else is around as "extremely harassing." In contrast, women view acts such as a man telling a woman that he sincerely likes her and would like to have coffee with her after work as signifying little or no harassment.[37]

I believe there are ample reasons to conclude that these findings about the profiles of sexual harassment victims, the gender differences in emotional reactions, and the importance of the attractiveness of the harasser all follow from the evolutionary psychology of human sexual strategies. Men more than women seek casual sex, and men's sexual over-perception bias leads them to infer sexual interest where none may exist. A concrete example of how this bias can lead to sexual harassment occurred when a supermarket chain implemented a new "superior customer service" program in which checkout clerks were instructed to make eye contact with and smile at customers.[38] Some women workers filed sexual harassment lawsuits, stating that their friendly conduct toward male customers frequently was misconstrued as flirtatiousness, leading to unwanted requests for dates, sexual propositions, and in some cases stalking. The supermarket subsequently changed its policy, which lowered the rate of sexual harassment.

Women sometimes feel hampered in their efforts to deter workplace harassment. When women do report sexual harassment, as occurred at Fox News when multiple women reported sexual harassment from host Bill O'Reilly, they are often silenced with monetary payoffs that carry strict legal nondisclosure agreements.[39] Women risk workplace retaliation. The psychologist Wendy Walsh hosted guest segments on Fox News and was then invited to dinner at a hotel restaurant with Bill O'Reilly. At the beginning, she reports, he was charming and discussed making her a more frequent contributor. After dinner, he invited her back to his hotel suite. When she declined, he became hostile: "All his charming-ness went away and he said the words, 'You can forget all the business advice I gave you, you're on your own.'"[40] Soon after, Fox News told her that her guest segments were being put on pause.

Because women who decline a man's overtures risk retaliation, some women attempt to deter unwanted overtures with "soft rejections" — phrases such as "I'm busy tonight" or "I have a boyfriend." Unfortunately, men sometimes construe these polite refusals as meaning "I would if I were not busy or if I didn't have a boyfriend, so perhaps sometime in the future." Sometimes these efforts to dissuade persistent men without incurring their wrath have the unintended consequence of keeping men's sexual hopes alive. When women say "I have a boyfriend," some men persist with comments like "Well, he isn't here" or "Break up with him." This forces harassment victims into a predicament in which they must choose between two potentially costly courses of action.

This broad-strokes account does not explain why some men are sexual harassers while it never even occurs to others to violate a woman's boundaries. Evolutionary science has identified two key predictors. The first is an inclination toward short-term mating, a general preference for casual sex over long-term mating with one partner. A study of 1,326 high school students in Norway revealed that a short-term sexual strategy proved to be the strongest predictor of which men perpetrate sexual harassment.[41] It was a better predictor than other potential predictors, such as the degree to which men had prior exposure to pornography, their endorsement of sexual stereotypes, their endorsement of myths about rape such as "Women ask for it by the way they act," and their acceptance of traditional gender roles. The fact that men inclined toward casual sex are precisely the men most likely to over-infer women's sexual interest suggests a dangerous combination of psychological features leading some men to become serial sexual harassers.

A second strong predictor of "which men" comes down to the personality trait of honesty-humility. Those low on this trait are considerably more likely to harass than those high on this trait.[42] These low-scoring men are prone to being interpersonally manipulative and dishonest in their interactions with others; are willing to cheat or steal to get ahead; believe they deserve special treatment;

and are notoriously lacking in empathy. In short, the subset of men most likely to sexually harass have all the hallmarks of the Dark Triad. Studies of more than twenty-five hundred Israeli community members showed that men who displayed high levels of narcissism, Machiavellianism, and psychopathy were far more likely to sexually harass than men who showed low levels of these traits.[43]

Most women in the workplace know that some men are especially prone to harass. Women often warn those new on the job to be careful not to be alone with certain men. Many women warned aspiring actresses about Harvey Weinstein's history of sexual predation, but Weinstein often circumvented these warnings by having a female assistant present during the beginning of the meetings, after which he would instruct the assistant to leave.

In summary, evolutionary science has identified at least some attributes of these men — those who are disposed to pursuing a short-term sexual strategy and those low in honesty-humility and high in Dark Triad traits. They perpetrate most of the harassment and are most likely to be serial harassers. These are also the men most likely to push the boundaries beyond mild forms of sexual harassment to the most severe forms of sexual coercion.

FROM STRANGER RAPE TO ACQUAINTANCE RAPE

A male approaches a female and begins to flirt. She is not aware that he has slipped her a drug that renders her unconscious. He then takes her body to his home, positions her correctly, and has sex with her while she is knocked out. If you've read about Bill Cosby, this may sound familiar. You may be surprised to learn that this is one of the mating strategies of the funnel-web spider, *Agelenopsis aperta*.[44] These male spiders have evolved a strategy of anesthetizing females who are reluctant to mate using a chemical cocktail.

Men obviously have not evolved adaptations to render women

unconscious. Some men, however, use an analogous strategy by plying their victims with evolutionarily novel drugs such as alcohol or Rohypnol. Harvey Weinstein almost invariably had a bottle of champagne on ice waiting in his hotel room, where he hosted his infamous "professional" meetings to discuss movies and movie deals. These meetings often ran late into the night, and Weinstein encouraged the women to stay overnight. One woman said: "I put on a tee shirt and shorts, and I went to bed. Suddenly, Harvey gets into bed naked next to me.... I pushed him away... and he tried to cajole at first, he said, 'Do you really want to make an enemy of me for five minutes of your time?' He just pushed and pushed. And he's huge. I weigh about a hundred pounds.... I don't know how to explain it. I thought if I just shut up, it will be over in a few minutes.... It's the collateral damage.... He just steals something."[45] The danger was real. Weinstein retaliated against women who refused him. This woman was faced with being raped, having her career ruined, or both.

An important feature of women's sexual psychology centers on the ability to correctly identify which men are most likely to be sexual predators. Are women accurate or miscalibrated in the modern world?

In a study of people's thoughts and fears about being murdered, Joshua Duntley and I asked five hundred people, roughly half women and half men, whether they had ever felt that their lives were in danger or felt that someone wanted to kill them. We were surprised by one set of findings we had not predicted: forty-six women (9 percent of the sample) described fearing that a man would rape them and then murder them. Of these women, 91 percent expressed a fear of being raped and then murdered by a total stranger; only 9 percent feared these outcomes from men they knew.

Here is one example. In response to the question "Who did you think might want to kill you?": "A stranger.... I was alone and he was staring at me. His eyes were intense. He started to walk faster toward me, speeding his pace. His eyes were cold and black. They seemed to bore right through me. His face was expressionless, blank. I thought he was going to drag me into the woods or into a dark alley and rape me, and then kill me with a knife. I ran for my life."

These findings surprised us for two reasons. First, the actual likelihood of rape victims being murdered is extremely low, at least in the United States. One scientist estimated it to be only one in ten thousand.[46] The vast majority of rape victims survive their attack, and most rapists use just enough threat or force to subdue their victims. Second, rape by strangers makes up a minority of overall rapes. Menacing strangers jumping out and attacking from a dark alley late at night are rare. Most sexual predators are people we know in settings we normally inhabit — at a party of a friend, at the apartment of a date, or in the sanctuary of our own homes.

These findings cry out for explanation. Why would women fear rape from strange men when most rapists are men they know? Why don't women show greater fear of rape from acquaintances since they make up the vast majority of perpetrators? And why would so many women fear that their rapist would murder them when the conditional probability of "if rape, then murder" is so small? In short, why do women appear at first glance to be miscalibrated in their rape fears?

One possible solution to this mystery is an *evolutionary mismatch*. Perhaps over human evolutionary history the rates of stranger rape and rape followed by murder were much higher than they are today. For thousands of years, small-group warfare often left a trail of female victims in its wake.

In the small-group living of our evolutionary past, with perhaps a few dozen to a few hundred in each group, people knew everyone in their group. There existed no within-group strangers. Strangers came from the outside, from other groups. Perhaps women's fears of strange men are well-calibrated defenses to an ancestral past in which strange men truly posed the greatest danger in the form of attacking bands of male warriors. It is possible that rape by acquaintances was rarer at the hands of ancestral men within one's group in part because no one lived alone. Groups were typically inhabited by one's close kin, such as mothers, aunts, sisters, fathers, and brothers, as well as female and male friends, all of whom could function as bodyguards and deter potential rapists. Although this of course requires some speculation, it is reasonable to believe that within-group sexual assault often would have brought swift and severe retribution from the victim's allies and kin, who would wound, ostracize, or kill the perpetrator. Social allies rarely tolerate sexual predators in their midst.

On the other hand, ancestral women who lacked bodyguards within the group may have been especially vulnerable to rape. Women with few or no kin around or women mated to a weak or cowardly man may not have been able to deter sexual violence from more aggressive formidable men. Reports of within-group rape may be found in numerous ethnographies of more traditional cultures, including the Gebusi of Papua New Guinea, the Yanomamo of Brazil, and the Tiwi of northern Australia.[47] And of course spousal rape undoubtedly occurred, although it may not have been categorized with that label.

In modern urban environments, women often move away from close kin. They change locales from the protective cocoon of a family to a geographically distant college for education or to large cities for employment. In most modern contexts, we are surrounded by strangers who create an evolutionarily unprecedented social world. In this novel environment, a friendly fraternity brother, a collegial classmate, or a casual internet date poses a far greater risk than a cruel, forbidding, unfeeling stranger. Women may also feel that since they know these men and have social relationships with some of them, they can better control the situation. They may have social currency and shared experiences they can leverage to appeal to the better angels of men's nature. They may trust the social contract implicit in a friendship. This sometimes results in misperceptions — they are assuming friendship when the men don't view their relationship that way. So women's stranger fears, supremely adaptive over the long stretch of human evolution, may be somewhat miscalibrated to modern social living. This is one way in which our highly functional ancestral psychology may not be well attuned to modern sexual dangers.

Another possibility is that women's fears of stranger rape are, in fact, supremely well calibrated to one set of dangers, even in the modern world. Because we encounter hundreds or thousands of strangers each year, even if a tiny percentage are potential rapists, a strong wariness of strange men could help women to avoid the catastrophic consequences of sexual assault. Stated differently, the modern rates of stranger rape may be low, relative to rates of acquaintance rape, precisely because women's stranger fears work so effectively — prompting defensive or evasive action to avoid becoming a victim. Because women are wary of strange men who stare at them, they manage to successfully avoid many of the situations that would put them in sexual danger.

These explanations, of course, are not inherently incompatible or mutually exclusive. And there are multiple possibilities. For instance, women simultaneously might be fairly effective at evading danger from strangers but possess insufficient defenses to protect them against the numerous acquaintances they encounter in the modern world. Sexual predators use internet dating sites such as Tinder, OkCupid, and Plenty of Fish to search for victims; registered sex offenders are sometimes not even screened by the sites.[48] Online communication with a sexual predator may give the illusion of knowing him, especially as people tend to attribute positive values to seemingly attractive online potential mates when there is missing information. Online dating is a modern technological form of mate attraction and mate selection for which humans have not had time to adequately adapt.

The modern world contains many evolutionarily novel circumstances. Colleges are inhabited by some men who ply women with alcohol at parties to disarm their defenses. Internet predators prey on vulnerable victims. Gender-integrated workplaces contain men who abuse power to sexually assault women in subordinate positions. The modern world, in short, has created evolutionarily novel hazards that allow predators to perpetrate sexual crimes.

THE CONTROVERSY OVER WHETHER MEN HAVE EVOLVED ADAPTATIONS FOR RAPE

So far, we have been exploring the possibility that sexual aggression may be the unwelcome byproduct of a more general adaptation in men to be on high alert for, and to be particularly pushy and persistent with, potential sex partners.[49] However, there is another theoretical possibility to consider. And that is the hypothesis that there may be male psychological adaptations specific to sexual coercion.

Rape is a charged topic about which emotions and moral revulsion run high. Evolution, too, is a fraught topic, especially when applied to human social behavior. And in the current cultural climate, the issue of sex differences has become a political hot button. Consequently, it should come as no surprise that combining these three factors — theories of rape that invoke evolved sex differences in sexual strategies — might generate controversy of epic proportions.

I hope that my treatment defuses some of the controversy, but not the urgency of addressing and eliminating these crimes. It is the job of scientists to dispassionately evaluate the merits of competing scientific hypotheses by rigorous standards of precision, parsimony, and evidence. For the goal of reducing or eliminating different forms of sexual coercion, understanding the variety of contributing causes is critical. To foreshadow my conclusion, after reviewing voluminous data, I will argue that there is no compelling evidence that humans have evolved specialized rape adaptations, and indeed some evidence flatly refutes specific rape adaptation hypotheses. At the same time, I will also argue that an evolutionary perspective of our evolved mating psychology is indispensable for understanding some of the root causes of sexual coercion, even if it is a highly unfortunate byproduct of other evolved psychological features. Improvements in the law and social norms, and dismantling patriarchal institutions, can change the social calculus such that the subset of men who desire to rape hold those desires in check as the social and legal punishments go up.

Evolutionary biologist Randy Thornhill and evolutionary anthropologist Craig Palmer outlined two competing theories of rape, one endorsed by each author. Thornhill proposed the theory that men have *evolved rape adaptations* — specialized psychological mechanisms for forcing sex on unwilling women as a reproductive strategy. Palmer proposed instead that *rape is a byproduct* of other evolved mechanisms, such as the male desire for sexual variety, a desire for low-cost consensual sex, a psychological sensitivity to sexual opportunities, and the general capacity of men to use physical aggression to achieve a wide variety of goals.

The *rape-as-adaptation* theory proposed several specialized adaptations that may have evolved in the male mind:

- Assessment of the vulnerability of potential rape victims (e.g., during warfare, or in non-warfare contexts where a woman lacks the protection of husband or kin)
- The mate-deprivation adaptation — a context-sensitive "switch" that motivates rape by men who lack sexual access to consenting partners, such as low-status males who cannot obtain mates through regular channels of consensual attraction
- A preference for fertile rape victims

- Sexual arousal in men to the use of force and to signs of female resistance to consensual sex
- Context-specific marital rape in circumstances where sperm competition might exist, such as when there is evidence or suspicion of female sexual infidelity

What does the scientific evidence show regarding these hypotheses? Let's first consider sexual arousal to the use of force, the hypothesis that has received the most scientific attention. One set of laboratory studies had men listen to tape recordings made by women reading first-person scripts.[50] In one condition, the women described in detail episodes of mutually consenting and satisfying sexual intercourse. In another, a woman's voice described a brutal rape in which the victims suffered fear and pain. While listening to the stories, the researchers measured men's sexual arousal. The scientists found that most men displayed high levels of sexual arousal to the consenting sex stories but low levels of sexual arousal to the rape story. When the male described in the story used physical violence such as slapping or hitting to enact the rape and the women described experiencing pain, sexual arousal in most men was dramatically reduced. Most men, apparently, are not sexually turned on by sexual violence toward women or victim suffering.

In contrast, convicted rapists in the study showed sexual arousal to both the consenting sex and nonconsenting rape depictions. The use of violence and hearing the women describe their suffering did not inhibit the men's sexual arousal at all. Professor Neil Malamuth and colleagues found similar results using samples of undergraduates who had no criminal record.[51] They found that arousal to stories of rape strongly discriminated between men who reported having used sexual coercion in the past versus men who reported no history of sexual coercion.

These findings do not support the hypothesis that most men have evolved a distinct rape psychology. Indeed, the finding that most men's sexual arousal is dramatically diminished when the rapist in the story uses violence or when the victim appears to experience pain actually contradicts the rape-adaptation hypothesis. On the other hand, the fact that men who report having used sexual violence or who have been convicted of sexual coercion in the past *do* show sexual arousal to rape depictions — and show no erectile diminution upon hearing the victim describe pain and suffering — suggests a distinct subgroup of men whose sexual psychology differs dramatically from that of most men. We will revisit the psychological features of this subgroup in greater detail shortly.

What does the scientific evidence reveal about the "mate-deprivation hypothesis," or what Professor Linda Mealey calls the "Making the Best of a Bad Situation" model? It is true that convicted rapists come disproportionately from lower socioeconomic groups. This seems at first glance to support the mate-deprivation hypothesis. Because women value economic resources in potential mates, men at the lower end of the economic spectrum may have more difficulty attracting women through normal courtship overtures. For men who lack the status, money, or other resources to attract women, coercion may represent a desperate alternative tactic, according to the mate-deprivation hypothesis. Men ignored or rejected by women because they lack the qualities for attracting desirable mates through honest courtship may develop hostility toward women, an attitude that short-circuits the normal empathic response and so promotes coercive sexual behavior.

The finding that *convicted* rapists come disproportionately from lower economic groups, however, is most plausibly explained by lower rates of reporting when rape is committed by men from higher social groups, combined with the greater ability of privileged men to evade arrest and conviction. Bill Cosby and Harvey Weinstein got away with their repeated sexual assaults for decades, in part because they were able to use their high-priced lawyers and wealth to silence victims with large monetary settlements and signed nondisclosure clauses.[52] Women sexually assaulted by high-status men are less likely to press charges, given the lower odds of being believed and obtaining justice. So the higher rates of convicted rapists coming from lower socioeconomic backgrounds is less revealing than it first appears.

Indeed, evolutionary psychologist Linda Mealey notes that college men who admit to behavior that legally meets the definition of rape tend to be more popular, have higher status, and have more consensual sex partners than other college men.[53] These "big men on campus" consider themselves to be so attractive and so high in value that they "feel no need to offer any personal attention, concern, or commitment to their potential sex partner(s).... They have already brought something extremely valuable to the bargaining table — so how could anyone possibly say 'no'?"[54] One contributing factor may be lack of empathy for the suffering of others. Several studies suggest that having higher social status is linked with lower levels of empathy; the wealthy, it turns out, tend to be more indifferent to other people's misery.[55] According to this hypothesis, it is popular high-status macho men, not mate-deprived low-status men, who are most likely to rape.

Supporting evidence comes from records of human history, in which kings, despots, princes, and priests have used their power to get away with many forms of sexual coercion.[56] Some rapes were actually codified in religious institutions. In his classic treatise on the history of marriage, Edvard Westermarck devoted an entire section to the religious leaders for whom "defloration is performed by the holy man," referring to the leader's right to have sexual intercourse with virgin brides before the betrothed man is permitted to do so.

In more modern times, the minority of Mormons who practice plural marriage continue to force young girls, some barely pubertal, into marriages and sexual engagement in which they have little or no choice — arguably a form of institutionalized rape.[57] The fact that some high-status Mormon leaders marry and sexually abuse large numbers of young women, while low-status men are often expelled from the group, illustrates that prestige and power are often linked to sexual coercion and monopolization. Religion, of course, does not have to enter the picture. This fact is exemplified by a quote from television celebrity and subsequently elected president Donald Trump, boasting about getting away with sexually assaulting women precisely because of his status: "When you're a star they let you do it.... You can do anything.... Grab them by the pussy."[58] The common practice in cultures with institutionalized slavery of wealthy slave owners raping female slaves also provides stark evidence against the notion that low-status sexually deprived men have some kind of monopoly on rape.

Psychological studies indeed provide direct evidence *against* the mate-deprivation hypothesis of rape. In a study of 156 heterosexual men, average age of twenty, evolutionary psychologist Martin Lalumière and his colleagues measured the use of sexual coercion with items such as: "Have you ever had sexual intercourse with a woman even though she didn't really want to because you used some degree of physical force?"[59] Separately, they measured mating success. Men who scored high on mating success *also* scored high on sexual aggression. Men who have had a lot of consensual sex partners were *more* likely to report also using force to obtain other sex partners. Other studies corroborate this finding. One study of a community sample of Canadian men found that the number of sex partners during adolescence, a key marker of short-term mating success, was one of the best predictors of sexual coercion.[60] Another study of people in rural South Africa found that wealthier higher-status men were more likely than poorer low-status men to rape women who were not their regular partners.[61]

A well-known example from Toronto, Canada, is Paul Bernardo.[62] Described as charming, good-looking, and educated, he was adept at pickup tactics and at attracting many women for consensual sex. He was also a serial rapist with more than a dozen victims. At age twenty-seven he married an attractive woman named Karla Homolka but continued his raping spree. Bernardo was eventually caught and convicted of sexual assault, torture, and murder and received a life sentence for his horrific crimes. The key point is that men who have success in attracting women are typically more likely to use coercion, and those who are less successful are less likely to use coercion. These findings refute the evolution-based mate-deprivation hypothesis of rape.

Although additional studies are needed, the available evidence leads to the conclusion that a simple version of the mate-deprivation theory of rape, in the form initially proposed by Thornhill and his colleagues, is almost certainly scientifically false. Although some disenfranchised men, those who bottom out in mate value in the eyes of women, sometimes rape, men with money, status, popularity, and power are more likely to be sexual predators.

One source of evidence sometimes used to support the evolutionary theory of specialized rape adaptations is that rape victims, like victims of sexual harassment, tend to be disproportionately concentrated among young, reproductive-age women. In one study of 10,315 individuals, women between the ages of sixteen and thirty-five were far more likely to be raped than women in any other age category.[63] Eighty-five percent of all rape victims are younger than thirty-six years old. Another study examined crimes of robbery in which the robbers opportunistically sexually assaulted victims who happened unexpectedly to be home — something that occurred in just under 2 percent of the 47,225 robberies studied.[64] Robbers who encountered a female while committing the robbery were most likely to rape her if she was between the ages of fifteen and twenty-nine, regardless of the age of the robber. Older and younger females were less likely to experience any form of sexual assault.

By way of comparison, victims of other crimes, such as aggravated assault and murder, show a markedly different age distribution. Women between forty and forty-nine, for example, are just as likely to suffer an aggravated assault as women between twenty and twenty-nine, but the older women are far less likely to be victims of rape. Indeed, the age distribution of rape victims corresponds well to the age distribution of women's fertility, in marked contrast to the age distribution of victims of other violent crimes unrelated to sex or mating.

That rapists disproportionately victimize young fertile women, however, is not evidence for or against the competing rape-as-byproduct and rape-as-adaptation theories. The finding can parsimoniously be explained by men's evolved attraction to observable cues to youth in consensual mating contexts. There is no evidence that this attraction is a rape-specific adaptation.

A conclusion reached by evolutionary anthropologist Donald Symons several decades ago seems apt today: "I do not believe that available data are even close to sufficient to warrant the conclusion that rape itself is a[n] adaptation in the human male."[65] Future scientific studies, of course, may alter this conclusion.

THE UNIVERSALITY CRITERION: ARE ALL MEN POTENTIAL RAPISTS?

In 1975, Susan Brownmiller argued that rape is "a conscious process of intimidation by which all men keep all women in a state of fear."[66] From an evolutionary perspective, the claim that all men are potential rapists requires identifying whether or not there exist universal features of men's sexual psychology that are designed specifically for forced sex. This is a formidable scientific challenge, as we saw in our discussion of rapists' tendency to target young women as victims. The possibility of men having a rape-specific psychology must be distinguished from men's sexual psychology in consensual contexts. No royal road paves the way to answering this question definitively, but several sources of evidence give us partial answers.

In one study, men were asked to imagine that they had the possibility of forcing sex on a woman against her will with no chance of getting caught, no chance that anyone would find out, no risk of disease, and no possibility of damage to their reputation. Roughly 35 percent indicated that there was some likelihood of rape under these conditions, although in most cases the likelihood was slight.[67] Another study using a similar method found that 27 percent of the men indicated some likelihood if there was no chance of getting caught.[68] Although these percentages are alarmingly high, they also indicate that most men are not potential rapists.

On the other hand, there are good reasons why the numbers of men who deny being willing to rape in a no-consequence condition may be underestimates. One is the obvious — rape is generally one of the most socially and morally condemned acts, along with incest and murder. Most people regard rape as a heinous abomination, although it's also true that some people do not regard an acquaintance forcing himself on a drunk date with the same degree of opprobrium as stranger rape. The key point, though, is that even those who think they might rape in a no-consequence circumstance may be reluctant to admit it on a questionnaire. A second reason to doubt these figures is that they rely on men being able to accurately imagine a scenario in which rape would have no chance of being discovered. It is unlikely that most men can correctly simulate this hypothetical scenario. Has there ever existed a circumstance in which the likelihoods of discovery and negative consequences were literally zero?

The closest approximation of a no-consequence context is that of warfare, especially when a victorious group has successfully defeated and killed their rival male combatants. Women in the defeated group become vulnerable and exploitable. The normal rule of law is absent. No statistics exist on the percentage of men who rape in these circumstances. The numbers vary tremendously based on social norms, approval or disapproval of war leaders, whether the crimes will be detected or undocumented, and many other factors.

What is known is that women victims are numerous and typically suffer devastating physical, psychological, and social consequences. In the Rwandan civil war between the Hutus and Tutsis, estimates are that roughly 354,440 women were raped.[69] As one scholar noted, "Almost every surviving Tutsi woman and adolescent girl was raped."[70] The pre-warfare rape rate for Rwanda is estimated to be 149 rapes per 100,000 women per year. During the hundred-day war, the rape rates were a staggering 220 times higher than the peacetime rates during the years preceding the war.[71] On the other hand, in other warfare contexts there are strict prohibitions, often coming from commanders, against rape. The hundred-day Rwandan civil war was almost certainly an outlier, one of the most extreme cases of rape in warfare in modern times. Still, it may not have been an outlier over human evolutionary history. Although we do not know the percentage of soldiers who rape in war, it is abundantly clear that the relatively consequence-free context of warfare disinhibits some men.

Another source of evidence bearing on the question of whether all men are potential rapists centers on men's *sexual fantasies*. Sexual fantasies are thoughts or mental images that often stimulate or enhance sexual arousal. Almost all people experience sexual fantasies of one sort or another.[72] Fantasies provide a window into our sexual psychology since they are typically private experiences less constrained by social taboos; the mind is freer to roam when no one can peer inside. Sexual fantasies stoke the fires of sexual arousal without requiring physical stimulation. Sexual fantasies commonly accompany sexual stimulation, though, whether through masturbation or during an actual sexual encounter. Perhaps most important, sexual fantasies sometimes motivate action. One function of fantasies is to motivate sex. Fantasizing about sexual coercion does not always lead to acts of sexual coercion, just as fantasizing about food does not always lead to acts of eating. People experience hundreds of fantasies for every one they act on. Nonetheless, acts of sexual coercion are often preceded by fantasies of sexual coercion.[73]

Do most men experience fantasies of rape? Scientific evidence on this question is sparse. A rare exception listed forty-six erotic sexual fantasies and had men report whether they experienced any of these fantasies.[74] At the high end, 84 percent of the men admitted to fantasizing about having sex with a woman other than their real-life sex partner. At the low end, 5 percent admitted to fantasies about sex with an animal. On sexual-force fantasies, slightly more than 33 percent admitted to fantasizing about a "scene in which you rape a woman."

Another study of sexual fantasies among 114 college men found that 36 percent of men agreed with the statement "I fantasize about raping a woman." A summary of six different studies found that 31 percent of men admitted to having experienced rape fantasies.[75] The similar percentages yielded by different studies, both of college men and community samples, lend credibility to the estimate of roughly a third of men reporting having rape fantasies.[76] If this estimate is taken at face value, most men do not have rape fantasies.

There are good reasons to suspect that these figures underestimate the actual numbers of men who experience ideation about sexual coercion. When the word "rape" was not used, one study found that 54 percent endorsed "I fantasize about forcing a woman to have sex" and 62 percent said that "it would be exciting to use force to subdue a woman." So the actual percentage of men who have forced-sex sexual fantasies may be higher than a third. Importantly, men who have these rape fantasies are also highly likely to report that they would actually commit rape if they knew that they would not be caught or punished.[77] This is a critical finding, supporting the idea that thoughts often precede actions. The thought-action link has long been recognized. It is contained in the Bible's Ten Commandments, which forbid not just adultery but also coveting one's neighbor's wife. Lustful thoughts lead to sexual deeds, at least some of the time, and provide the motivational impetus for them.

On the other hand, 90 percent of men said "I don't understand how a man could possibly rape a woman" and 77 percent said "If I knew my roommate raped a woman, I would report him."[78] Several interpretations might explain these findings when they are juxtaposed with the findings from the same men who report fantasizing about forcing a woman to have sex. One is that men entertain thoughts of forced sex but would not act on those thoughts. Another is a double standard of self versus other. Perhaps forced sex may be morally condemned when perpetrated by others but not when perpetrated by oneself. A third explanation stems from the word "rape." Some men admit to forcing sex on a woman without her consent but do not label the act as rape. So the high moral condemnation of others who rape resides in men's minds along with the idea that forced sex does not qualify as rape. And we cannot forget the power of rationalization. Some men erroneously believe that their victims "really wanted it."

Fantasies are one thing, actions another. How many men actually force women to have sex without their consent and against their will? One source of evidence comes from studies that explicitly avoid using the word "rape" but instead ask men: "Have you ever had sexual intercourse with an adult when they didn't want to because you used or threatened to use physical force?" A study of 1,882 American men found that 120, or 6.4 percent, admitted that they had.[79] Of these, about two-thirds were repeat rapists, averaging 5.8 admitted rapes. This sample consisted not of convicted rapists but of college students attending a midsize urban commuter university. Other studies have found that between 6 and 15 percent of college males admit to rape or attempted rape as long as the word "rape" is not included in the description.[80]

A large cross-cultural study used a similar method, asking 10,178 men from six different countries these two key questions: have you "forced a woman who was not your wife or girlfriend at the time to have sex" or "had sex with a woman who was too drunk or drugged to indicate whether she wanted it"?[81] The percentage of men who admitted having raped a non-partner was 4.3 percent in Bangladesh, 6.2 percent in Sri Lanka, 8.1 percent in China, 8.3 percent in Cambodia, and 12.8 percent in Indonesia. The most extreme country in the study was Papua New Guinea, which logged an astonishing 40.7 percent of men admitting to the act of rape when that specific word was not used. Most men are not rapists when evaluated by the criterion of actual behavior, but a subset of men clearly are rapists. Can this subset be identified?

THE PSYCHOLOGICAL CHARACTERISTICS OF MEN MOST LIKELY TO RAPE

Only a subset of men are particularly prone to rape. Men who use coercion to get sex have been shown to exhibit a distinct set of characteristics. They tend to be hostile toward women, endorse the myth that women secretly want to be raped, and show a personality profile marked by impulsiveness, aggressiveness, low empathy, hypermasculinity, and narcissism. This constellation is particularly dangerous when combined with a high degree of sexual promiscuity.[82] Rapists have the dangerous combination of high levels of Dark Triad traits — psychopathy, narcissism, and Machiavellianism — and a disposition to pursue a strategy of casual rather than committed mating.

It is worth examining in greater detail this underlying psychology; doing so provides important clues about the psychological machinery that prompts some men to override social norms, moral principles, and legal prohibitions against rape. Most people experience *empathy* and compassion for victims of violence. Empathy is an emotion that prevents potential perpetrators from inflicting violence. Experiencing pain at a victim's suffering deters most men from causing that suffering. Severing the empathy circuit frees a subset of men to act on their desires and renders the rape victim's suffering irrelevant. Consider the words of this rapist, a forty-nine-year-old male: "I wasn't concerned with her fear. I was only concerned with her body.... Her feelings didn't mean anything to me at all."[83] In marked contrast, empathy likely motivated two students to intervene and stop Stanford student Brock Turner from continuing to sexually assault an unconscious woman behind a dumpster and hold him until the police arrived.[84]

Hostility is another key quality. Hostile men feel wronged by others. Men hostile to women feel that they have been wronged by women and that women cannot be trusted. They see themselves as victims, either of women who spurn their advances or of women whose sexual attractiveness inflames their sexual urges. Rage can fuel forced sex. Rather than feeling empathy for the victims' suffering, men angry at women may actually enjoy inflicting pain. Men who have a high level of *hostile masculinity* endorse statements such as "A lot of women seem to get pleasure out of putting men down," "A man's got to show a woman who's boss right from the start," and "In a dating relationship a woman is largely out to take advantage of a man."[85] Hostile men see the sexes as fundamentally at odds with each other in a zero-sum game.

Narcissism is a personality trait that studies sometimes link with forced sex.[86] Three elements of narcissism are especially relevant — a grandiose sense of entitlement in which the individual believes that their desires should be fulfilled just because of who they are; interpersonal exploitativeness, in which others are manipulated or coerced; and a lack of empathy. The narcissist's exaggerated sense of entitlement leads him to believe, compared to men who score lower on this trait, that he deserves sexual favors from women he desires. Narcissistic men are more prone than other men to see a sexual refusal as a personal affront and react with hostility to what they see as a rejection. Narcissists tend to have grandiose views of their sexual prowess and sexual superiority, so a woman who rejects their sexual advances is interpreted as impugning their manhood, evoking rage. This psychological constellation prompts some men to rectify the apparent offense of a woman's sexual refusal by overriding it.

These traits prove especially dangerous when combined with a disposition to pursue a *short-term mating strategy*.[87] Psychologists typically measure three components of a short-term mating strategy: behavior, attitudes, and desire. Behavior is measured by items evaluating number of past sex partners and by items such as "With how many sex partners have you had sex on one and only one occasion?" Attitudes are measured by items such as "Sex without love is OK." Desire is measured by statements such as "How often do you fantasize about having sex with someone other than your current dating partner?" and "I can imagine myself being comfortable and enjoying 'casual' sex with different partners."[88]

Men with a short-term mating strategy, in brief, tend to have a high sex drive. When these high-libido men are low in Dark Triad personality traits, they implement their strategy through tactics that include honest courtship, charm, and seduction. When combined with psychopathy, a low level of empathy, narcissism, and hostile masculinity, high-libido men implement a short-term strategy through deception, threats, force, and disabling victims through alcohol and drugs. This toxic mix exposes women to sexual predation.

These men are also prone to commit multiple rapes. To get a

sense of the magnitude of this problem, consider the fact that for each crime a rapist is convicted of committing, there are many more that have gone undetected. When one study gave convicted rapists assurances of confidentiality, it found that 126 rapists admitted to committing sexual violence against 882 victims.[89] In another study of 37 rapists who had been charged with 66 offenses, or roughly 1.8 victims on average, the rapists admitted to 433 rapes when assured of confidentiality, with an average of more than 11 victims per rapist.[90] Yet another study of 120 men whose self-reported behavior met the legal criteria for rape but who had *not* been charged with, or convicted of, any of these incidents found that victim number averaged roughly 4 rapes each, totaling 483 rapes.[91] Most of these men, roughly 81 percent, raped women who were incapacitated due to alcohol or drugs. The others used threats of force or actual force to subdue their victims. A full 38 percent of these men also admitted to the nonsexual battering of an intimate partner, highlighting a link between intimate partner violence and rape.[92]

Importantly, these figures reveal a disturbing number of undetected rapes and undetected rapists. A small percentage of men leave a long trail of victims in their wake, whether criminally charged or not. The fact that most rapists are serial rather than onetime rapists partially resolves an apparent paradox noted in the scientific literature — that a large number of women report being raped, but a much smaller number of men report committing rape.

MARITAL AND PARTNER RAPE

Most scholars distinguish spousal rape as a conceptually separate kind of rape. Laws have traditionally separated it, going back to seventeenth-century English law: "The husband cannot be guilty of a rape committed by himself upon his lawful wife, for by their mutual matrimonial consent and contract the wife hath given up herself in this kind unto her husband, which she cannot retract."[93] In the 1970s, marital rape was legal in all fifty US states. By July 5, 1993, all fifty states had laws on the books criminalizing it. Some countries around the world, from Albania to Yemen, have continued the tradition of exempting marital rape from criminal statutes. Nonetheless, the trend worldwide is clearly improving. Criminalization of marital rape started in Australia in 1981, in Brazil in 2005, in Albania in 2012, and in Barbados in 2016.[94] The cultural progression of laws and attitudes surrounding this form of institutionalized sexual assault is rapid and moving in only one direction. Despite changes in the legal codes, some people still perceive marital rape to be less of an infringement on women's rights and less psychologically traumatic than stranger or acquaintance rape.[95] Whether changes in laws have led to commensurate changes in actual rates of rape remains unclear.

Estimates of the rates of marital rape are problematic, of course, in part due to the low rates of criminal charges filed and the mistaken belief that forced sex within a marriage can't constitute legal rape. For example, some women describe their husbands as threatening harm and using force to have sex against their will but do not categorize the acts as rape. Despite these difficulties of obtaining true rates, studies put the rate between 5.7 and nearly 16 percent of all married couples.[96] One telephone poll of 1,108 women, for example, found that 13 percent reported that their husband had used force or the threat of force to obtain sex.[97]

An evolutionary perspective sheds light on why partner rape, what Professor Linda Mealey calls "The Mate-Guarding Model," should be distinguished from other forms. The circumstances in which it occurs differ dramatically from those of acquaintance or stranger rape. Misperception of sexual interest, for example, is a circumstance conducive to acquaintance rape but is unlikely to play a key role in spousal rape. A distinguishing feature is that marital rape is more likely when the husband is concerned about a suspected or actual sexual infidelity. This feature is not relevant to acquaintance or stranger rape. Similar patterns are seen in other species. American black ducks, mallards, and shrikes also attempt forced copulation on their regular mates when they perceive cues that another male may have copulated with her.[98]

Married women also experience a dramatically elevated risk of rape during or immediately following a breakup. This finding supports Mealey's mate-guarding model when it is linked with the finding that some women use affairs in order to facilitate exiting from a bad relationship and transitioning to a better mate.[99] Men often view a spouse who is leaving as irreplaceable. Forced sex in these circumstances may reflect the urge to maintain control over what, from an evolutionary perspective, is a desirable woman who might be lost forever. This does not imply that men have evolved adaptations specifically for marital rape. Rather, this reflects men's more general psychological mindset of gaining sexual access and jealously striving to retain that access by whatever means they feel entitled to implement.[100]

As in other types of rape, men are more likely to rape younger compared to older wives. Like other age-of-victim findings, no specialized rape adaptation is needed to explain them. A large age disparity, in which the husband is ten or more years older, puts married women at greater risk.[101] Married women who are more economically dependent on their spouses experience greater risk of spousal rape compared to women who have their own independent resources.[102] Women with more resources are freer to leave bad marriages. Women lacking resources often feel stuck in sexually abusive relationships.

Husbands who score high on hostile masculinity and an orientation toward impersonal sex are more likely to use force or the threat of force against their wives. A study of 1,370 South African men inhabiting rural villages found that men who rape intimate partners are more likely to have had a larger number of sex partners.[103] Precisely the same risk factors — the combination of Dark Triad traits and an orientation toward short-term sex — are characteristic of rapists inside and outside of marriage.[104]

THE CURSE OF SEXUAL COERCION

Although there is little evidence that human males have evolved specialized adaptations for rape, and some evidence against specific versions of rape-adaptation theory, sexual coercion cannot be understood without considering scientifically documented features of men's mating psychology. The key features of this psychology that cause so many problems when it comes to coercion include:

- Men's strong attraction to young women, even as men get older and occupy positions of power
- Men's attentional adhesion to attractive women
- Men's brain reward circuit in the nucleus accumbens, which becomes activated when men view attractive women
- Men's sexual arousal to visual information, which abounds in modern environments
- Men's high sex drive and the powerful emotion of lust
- Men's desire for sexual variety, which leads to attraction to novel women, even when men have existing sexual partners
- Men's greater motivation to pursue a short-term mating strategy
- Men's willingness to have impersonal sex with no emotional attachment
- A sexual over-perception bias that leads some men to believe that sexual interest is present when it is not

- A link in men's minds between power and sex that leads to an unwarranted sense of sexual entitlement, especially among men prone to sexual harassment
- A short-circuiting of normal empathic emotions for victims of sexual coercion in some circumstances
- The presence of Dark Triad traits — psychopathy, Machiavellianism, and narcissism
- Low scores on the personality trait of honesty-humility, which is linked with aggression and lack of empathy
- The general willingness of men to use violence or threats of violence to achieve a variety of goals, including obtaining tools, territory, money, food, and sex

Four facts loom large in the chasm between men's and women's minds. First, men are sexually attracted to many women who experience no attraction to them. The reverse is also true sometimes, although it is less frequent. Second, men want to have sex with many women who do not want to have sex with them. Third, because people often infer the sexual desires and motives of others by extrapolating from their own minds, men profoundly fail to understand that women's sexual psychology differs from their own. They have trouble conceiving that the woman whom they find so arousing may not hold a glimmer of attraction to them. Fourth, some men are prone to use violence and threats of harm to get what they want. To be clear, though, these features of male psychology are not inevitably or invariantly expressed. "As soon as we recognize that there is nothing morally commendable about the products of evolution," Harvard psychologist Steven Pinker observes, "we can describe human psychology honestly, without the fear that identifying a 'natural' trait is the same as condoning it."[105]

Importantly, the expression or inhibition of dangerous elements of men's sexual psychology depends on key personal and social conditions. The social context of sex ratio in the mating pool provides a vivid example. When there is a surplus of men compared to women, rape rates go up. When there is a surplus of women, rape rates go down.[106] This is especially pronounced in some areas of India and China in which a preference for sons can lead to selective abortion and infanticide of females, producing a large surplus of men. Eliminating these male-biased practices might reduce rape rates.

Other factors that influence the expression or suppression of dangerous features of male sexual psychology include social norms that render sexual harassment acceptable, as in the "boys will be boys" cliché, versus policies that render it a fireable offense; laws that do or do not make spousal rape a crime; the Dark Triad and a strategy of casual sex that combine to render some men much more likely to be sexual coercers. A diminution of patriarchal ideology, stronger enforceable laws, greater police sensitivity to victims of sex crimes, and a more educated populace on these issues can have salutary effects in curtailing male sexual coercion.[107] An evolutionary perspective is indispensable in understanding some of the otherwise puzzling patterns in sexual coercion. Ideally, this knowledge can help to promote its total elimination.

This chapter has focused primarily on men's sexual psychology, the elements of men's minds that combine to incline some to perpetrate sexual violence. Men's sexual psychology, however, did not evolve in a vacuum. The fundamental principle of sexual coevolution requires that we must simultaneously understand women's sexual psychology. That psychology is a core part of the social matrix within which male sexual psychology evolved.

Whatever are the multiple causes of sexual violence, there is no doubt that it is profoundly harmful to victims. Costs create selection pressure for defenses to prevent becoming a victim to those harms. Although it is obvious that victims of sexual coercion suffer psychological trauma, few scientists stop to ask why this form of violence exceeds all others in psychological damage. Counterstrategies to defend against sexual predators are central to the sexual coevolutionary process. And they start with the first law of human mating — women's choice about when, where, with whom, and under which circumstances they consent to sex, a topic to which we now turn.

CHAPTER 8

DEFENDING AGAINST SEXUAL COERCION

> Most men fear getting laughed at or humiliated by a romantic prospect while most women fear rape and death.
>
> — Gavin de Becker

THE GREAT VARIABILITY ACROSS CULTURES and subcultures in rates of sexual coercion offers hope for successful intervention. Some cultures contain conditions more conducive to rape than other cultures — different social norms, different laws, different enforcement of laws, size and anonymity of living conditions, proximity to close kin or bodyguards, ratio of men to women in the population, and many others. Identifying the specific conditions that comprise what is sometimes termed "rape culture" is critical for eliminating rape. So is identifying the characteristics of men most likely to perpetrate it and women's defenses to prevent it.

Everyone except the most oblivious knows that rape is psychologically traumatizing to victims. An evolutionary perspective illuminates with greater precision why women experience rape as so harmful. The emotional experience of psychological trauma helps us to mark and monitor events that are harmful to us and our loved ones — injury due to attack, an illness that befalls a loved one, the pain of a partner's betrayal. It should not surprise us, then, to find that sexual coercion has imposed substantial injuries on women. Before turning to these many damages and women's defenses against them, an evolutionary lens compels us to ask a key prior question: Is rape a relatively recent occurrence, or has it been a recurrent hazard afflicting women over deep time?

SEXUAL COERCION THROUGHOUT HUMAN HISTORY

If the written historical record, cross-cultural ethnographic evidence, bioarcheological findings, and molecular genetic evidence suggest that rape has occurred throughout human history, it would defy logic if evolution by selection had not fashioned defenses in women to prevent its occurrence. Importantly, this is a separate issue from the question of whether men have evolved adaptations to rape. Women could have evolved anti-rape adaptations, in principle, even if rape has been entirely a nonadaptive or even maladaptive byproduct of other features of male psychology such as aggression proneness and the male sexual over-perception bias. Humans have an evolved fear of falling from heights such as from tall trees and towering cliffs, for example, even though trees and cliffs did not develop their height to inflict damage on humans. Defenses can evolve to recurrent hazards even in the absence of evolved offenses, and we concluded in the previous chapter that there is no currently compelling evidence that men have evolved rape adaptations.

We can never determine with absolute certainty whether rape was frequent enough historically to have forged female anti-rape defenses. But enough evidence has accumulated to allow us to make an educated guess. Written history dating back to the Bible brims with episodes of rape, and even specifications by religious leaders about the conditions under which men can sexually assault women. Most of these conditions involve periods of warfare.

For example, the Sages of the Talmud, codified by Maimonides, provide this injunction:

> A soldier in the invading army may, if overpowered by passion, cohabit with a captive woman...[but] he is forbidden to cohabit a second time before he marries her....Coition with her is permitted only at the time when she is taken captive...he must not force her in the open field of battle...that is, he shall take her to a private place and cohabit with her.[1]

Written records outside of religious texts that effectively give license to rape reveal similar themes, such as these by the feared warlord Genghis Khan (1158–1227), relishing the rape of conquered groups: "The greatest pleasure is to vanquish your enemies, to chase them before you, to rob them of their wealth, to see their near and dear bathed in tears, to ride their horses and sleep on the white bellies of their wives and daughters."[2]

No systematic studies have been conducted on the occurrence and frequency of rape among traditional societies across cultures. Moreover, since rape is typically carried out beyond the view of anthropologists studying particular cultures, ethnographic reports undoubtedly reflect an impoverished record and underestimate its occurrence. Several scholars have attempted to scour the ethnographic evidence, however, and they find that rape is indeed present in many traditional societies. From the Amazonian jungle of Brazil to the more peaceful !Kung San of Botswana, ethnographies contain references to rape ranging from mentioning it in passing to providing detailed descriptions. The Semai of central Malaysia, for example, were frequently victimized by Malay raiders, who would ambush them, kill the men, and take the women by force.[3] The Amazonian peoples studied by Thomas Gregor have special words for both rape — *antapai* — and gang rape — *aintyawakakinapai*.[4] The Yanomamo of Brazil state that the primary reason they attack neighboring groups is to capture women, or to recapture women who have been taken from them in previous raids.[5] The ethnographic evidence, in short, is rife with reports of rape.

Rape in warfare is disturbingly common, as Susan Brownmiller documented extensively in her 1975 book *Against Our Will*.[6] Her findings have been corroborated by other researchers and historians. *The Rape of Nanking*[7] documented the thousands of rapes and murders of Chinese women by invading Japanese men during World War II. Less publicized was the frequent rape of Jewish women by German soldiers during the World War II Holocaust.[8] *Rape Warfare*[9] provided similar documentation of rapes of women in Bosnia and Herzegovina and in Croatia during the 1992–1995 wars. These crimes also occurred in the Democratic Republic of Congo throughout the Second Congo War (1998–2003) and are still occurring in the ongoing postwar conflicts in the same region.[10]

Some scholars claim to have identified cultures where rape is entirely absent, such as among the Mbuti, the Yap, the Arapesh, the Huichol, and the Mataco.[11] These claims, however, are typically contradicted by the actual ethnographies. Among the Mbuti, for example, men are said to require permission to have sex with a woman. Nonetheless, "the men say that once they lie down with a girl, if they want her they take her by surprise when petting her, and force her to their will."[12] Among the Yap, ethnographers report that when battling other villages, women are captured for sexual purposes and that a Yap man "sometimes tries to obtain by force what cannot be attained in an amicable way."[13] Among the Arapesh, who Margaret Mead claimed did not commit rape, ethnographers report instances in which Arapesh men forcibly abduct women for the purpose of sex. Among the Mataco, another culture coded as lacking in rape, "Wars are often waged for the very purpose of stealing young women who are afterwards married by the victors."[14]

No firm conclusions can be drawn about the prevalence of rape in traditional cultures. Rates undoubtedly vary dramatically across cultures from rare to common, and identifying sources of cultural variation can provide critical clues to eliminating rape. We can firmly conclude, however, that rape is reported with enough regularity across scores of ethnographies to suggest that it posed, and continues to pose, an important problem for some women some of the time.

Paleontological evidence also reveals an intriguing fact that bears on the subject of prehistoric rape. Caches of bones depicting the tragic aftermath of war reveal mostly male skeletons with arrow tips lodged in rib cages, blunt force trauma, and cranial wounds that correspond in size and shape to the weapons found in the vicinity.[15] Easy to overlook, however, is the fact that these skeletal graves often contain a predominance of males and a striking absence of skeletons of reproductive-age females. This pattern is seen in prehistoric Australia, the Crow Creek burial pits in South Dakota from the year AD 1325, Koger's Island in the American Southeast, the American Southwest, Mexico, South America, Easter Island, Scandinavia, and Western Europe. The notable dearth of young women among the skeletal remains of the burial sites provides evidence that dovetails with the written and cross-cultural evidence that reproductive-age women were captured forcibly during warfare for sexual or marital ends.

Molecular genetic studies also provide corroborating evidence. From what we can tell, the feared warlord Genghis Khan may have been the most prolific rapist in human history. He united previously disparate tribes to wreak havoc over a twenty-year time span, from roughly AD 1206 to AD 1227. Khan conquered groups and created an empire that extended some six thousand kilometers from the edge of Japan in the east to the Caspian Sea in the west.[16] He commanded loyalty in part by sharing resources from the conquered tribes, including the women left vulnerable by the slaughter of their male defenders. Genghis Khan apparently kept the most attractive women for himself and his sons, housing them in harems. Genetic analysis of the DNA of central Asians reveals that roughly 8 percent, or roughly 16 million men alive today, have virtually identical Y chromosomes, a signature that reveals that they all came from the same man, almost surely Genghis Khan.[17] Genetic analyses reveal similar, although less prolific, patterns in Ireland (King Niall).[18] The Vikings from Scandinavia invaded and vanquished Irish male populations, leaving kings, descendants, and Y chromosomes in their wake. The Scandinavian Vikings also overran Iceland with similar genetic effects.[19]

In Europe, Asia, and Africa, the male population, but not the female population, plummeted to roughly one-twentieth of its former size from roughly seven thousand to five thousand years ago — a dramatic evolutionary bottleneck.[20] Tribal clashes involving clubs, arrows, and axes were almost surely responsible. The dramatic drop in male population is inferred from the sharp decline in the diversity of Y chromosomes, which are passed down exclusively from fathers to sons, without any commensurate drop in diversity of mitochondrial DNA, which is passed down exclusively from mothers to daughters. It is impossible to escape the conclusion that the surviving women of the vanquished groups became the sexual captives of the victorious males.

In short, much evidence converges on the conclusion that women going back hundreds of thousands of years have suffered from different forms of sexual coercion — from opportunistic rape to the capture and de facto sexual enslavement consequent to war. If it has occurred throughout human history — and the written record, cross-cultural ethnographic evidence, bioarcheological findings, and molecular genetic evidence suggest that it has — we must now turn to the question of its catastrophic costs to women.

THE MANY HARMS OF RAPE

In the first study my lab carried out on conflict between the sexes, we were able to identify 147 actions someone could perform that might irritate, anger, annoy, or upset a member of the opposite sex.[21] These ranged from the relatively minor, such as "He left the toilet seat up," to the more serious, such as "He hit me when he was upset." Then we asked panels of women and men to evaluate how upsetting each act would be to them personally. We also asked separate panels of third-party judges to assess how upsetting they thought each act would be when experienced by a woman and by a man.

Having a partner who was sexually unfaithful, verbally abusive, or self-centered was equally upsetting to women and men. But there was one enormous sex difference: sexual aggression. This cluster included acts such as someone touching your body without your permission, demanding sexual relations, threatening you verbally in order to have sex, and forcing you to have sex. Women were far more upset about being victimized by sexual aggression than by any of the other fourteen clusters we identified, including being physically abused by a romantic partner, which of course was also very upsetting. Most women rated being the victim of sexual aggression as "extremely upsetting," the highest rating on the 7-point scale. The average man was far less bothered by being sexually victimized in these ways by a woman, rating them as only modestly bothersome.

In the third-party ratings — how upset would someone else be by sexual aggression — both sexes viewed acts of sexual aggression as extremely upsetting to women, and more upsetting to women than to men. However, there was a critical psychological disconnect: although men recognized that sexual aggression would be more upsetting to women than to men, they dramatically underestimated *how upsetting* women would actually find these acts. This psychological disconnect is illustrated by the then Texas political candidate who declared, "Rape is kinda like the weather. If it's inevitable, relax and enjoy it."[22] The comment infuriated many and he subsequently apologized, claiming that it was a joke. But the mere fact that he would treat the issue so lightly highlights the immense gap in men's ability to understand women's fear of rape and the trauma that ensues from its occurrence. Fortunately, most men are not as clueless as this then political candidate. Men whose partners, sisters, or female friends have been raped have greater empathic understanding, and of course some men have been victimized themselves. But when it comes to rape, the chasm in cross-sex mind reading is profound.

These findings were presciently predicted by feminist scholars who made a major contribution to the understanding of rape from the victim's perspective.[23] Diane Russell interviewed many rape victims, discovering common themes in their experiences. Several conclusions are clear. Contrary to what some men believe, women do not want to be forced into sex. Nor do they experience rape as a sexual act. The idea that it could be enjoyable is ludicrous. Most women view rape as an act of violence. Rape victims commonly feel intense rage at being violated, sinking into deep shame, humiliation, and depression. Many victims suffer from post-traumatic stress, which can last months and even years, a topic discussed in detail later in this chapter.

Perhaps these observations seem obvious, but an evolutionary perspective poses a deeper question that is rarely asked: Why? What accounts for the fact that women experience rape as more traumatic than any other crime that could afflict them? The answer starts with the critical fitness impacts of disrupting a fundamental feature of female mating strategies across the animal kingdom — female choice.

A woman's ability to decide when, where, and with whom to have sex has always been profoundly important and beneficial.

Good mating decisions come with many benefits. Over evolutionary history, a desirable mate could provide protection for a woman's children, food during bitter winters or when pregnancy limits her mobility, access to social allies such as in-laws, and good-quality genes such as those promoting a healthy immune system. Importantly, according to some evolutionary scientists such as Barbara Smuts, Sarah Mesnick, Linda Mealey, and Margo Wilson, protection from sexual exploitation by other men is a key benefit of choosing a high-value mate. Given all the benefits of exercising mate choice, it should be clear why circumventing it through sexual coercion is so harmful.

Women's preferential mate choice favors men with attributes that augur well for receiving these benefits. Traits like honesty, industriousness, dependability, bravery, physical strength, intelligence, the ability and willingness to invest in her and her children, and the ability to protect them from harm all bode well for a woman's offspring.[24] Rape can deprive a woman of these key benefits normally obtained from a carefully chosen consensual romantic relationship. Sexual aggression interferes with this most fundamental female mating strategy and inflicts massive damage in evolutionary currencies.

A rape puts women at risk of physical harm, since women can suffer bodily injury in the course of the crime. The bodily harm is compounded by an increased risk of contracting a sexually transmitted infection from the perpetrator. Injury in ancestral environments, which lacked antibiotics and other forms of modern medicine, would have increased the risk of infertility and even death. If a rape victim becomes pregnant with the rapist's child, rather than the child of a man of her own choosing, she risks missing out on the benefits afforded by female choice. She risks raising the child without the help of an investing mate. Her perceived desirability also suffers: most men view women burdened with another man's child as a liability, not a benefit, on the mating market. Of course, many women with children sired by other men do succeed in attracting an investing mate. If they succeed, however, their children will become stepchildren. Stepchildren suffer rates of psychological abuse, physical abuse, and even murder many times greater than children growing up in intact families with two biological parents.[25] Being a stepchild directly reduces survival probability, diminishing the biological mother's potential reproductive success.

Children resulting from rape sometimes suffer rejection by their own families. As one Rwandan mother put it: "I am wondering who will bring her up after my death. My aunt who survived the genocide doesn't like my child as I do. She said that it will be an eternal torture to bring up an Interahamwe's child as she will remind her how the Interahamwe have decimated our family."[26]

In addition to these enduring tragedies, some victims of rape and their children suffer severe social stigma, social isolation, and shame from their communities.[27] One rape survivor of the Rwandan genocide observed, "It is a big problem to be known as a rape survivor in the community. They didn't respect you, they isolated you, people said that we were no different from prostitutes because we accepted having sex with any man who wanted to have sex with us during the genocide."[28] Another woman said, "I sometimes feel like committing suicide. I feel so ashamed because everyone in the neighborhood knows that I was raped.... People mock me in the neighborhood."[29] From an evolutionary perspective, acquiring social stigma can be catastrophic, with cascading consequences ranging from losing allies to being last in line during food shortages.

Social isolation can be self-imposed, as some rape victims withdraw. One woman in our study of college rape victims captured it this way: "I basically became a hermit and refused to go out to party or socialize. This also included skipping a great deal of class and just moping around my dorm. I would stay on my computer...and only venture out at night for late-night snacks and talks with a few close friends. Many of my friends didn't understand my sudden hermitic lifestyle so I have lost quite a few."[30] Social isolation, whether self-imposed or due to ostracism, can cause a victim to spiral downward into depression and despair.

The rape of a daughter or sister sometimes can bring shame upon the entire family. In some cultures, the community expects the kin to punish or even kill the rape victim lest the entire family suffer shame and shunning.[31] Parents and siblings sometimes distance themselves from the victim, weakening her most valuable line of protection. Friends, likewise, sometimes detach themselves for fear that they too will be tainted by association. Damage to the victim's kin relations, friendships, and social alliances is exceedingly harmful from an evolutionary perspective. Historically, these social alliances were always critical components of the survival strategies of our ancestral mothers; no one could survive long on their own.

Compounding social rejection from their families and communities, the severe ramifications continue throughout life, decreasing the desirability of the rape victims as well as their children when they mature to adulthood. As one rape victim put it, "The experience of being raped had changed my life. Because I didn't feel as a girl because I am not a virgin but I am not a woman either because I have no husband and I know no one will agree to marry a girl who is not a [virgin]."[32] Some cultures, of course, will inflict these social costs more than others, and there are movements in European and North American countries to reduce the stigma attached to rape victims. In some cultures, however, being raped makes a woman unmarriageable.

Women victims also sometimes suffer from decreased self-perceptions of desirability, undermining their aspiration to attract a high-quality long-term mate. As one college-age woman in our study of rape described it, "My self-esteem plummeted after the sexual victimization. I was depressed and didn't think myself worthy of dating other guys. I was scared to get too close and intimate with other guys, and I was really beating myself up over the fact that I didn't fight back more and try to resist the sexual victimization. I thought I was weak, stupid, and naive. I didn't think of myself [as] attractive or worthy of care from other people. I just didn't have confidence in myself anymore" (twenty-three years old, raped at age eighteen).[33] This quote highlights how closely intertwined self-esteem is with self-perceived sense of worth as a desirable mate.

The trauma extends to married women and those in committed mateships. A mated woman raped by another man risks losing her current romantic partner. He may perceive her to be less desirable, unfairly viewing her as "damaged goods." He may worry about harm to his own social reputation, being perceived as being cuckolded. He may regard the rape as a form of infidelity, especially if he has the slightest suspicion that there might have been some element of consent or if the victim does not show injuries consistent with the violent resistance that he expects a rape victim to display. These harms are well captured by a woman in our study of rape who described the aftermath of her rape on her relationship in this way: "It ruined it. [My partner] couldn't get it out of his head that it was my fault. He started criticizing my behavior and what I wore. Basically, 'I think how you act and how you dress gives guys the idea that you want to have sex with them.' It wasn't, 'You look beautiful in that,' it was, 'Guys are going to think you want sex.' Our relationship ended because of stuff like that" (twenty-one years old, raped at age nineteen).[34] Romantic breakups and divorces climb sharply in the aftermath of rape.

These social costs and the risk of ostracism by their own families are undoubtedly some of the reasons prompting women to conceal their victimization, producing the massive underreporting rate for these crimes. Who can blame a woman for concealing a rape when revealing it compounds its costs so exorbitantly? Why let others know when doing so can amplify the trauma?

Yet keeping it hidden also produces unintended harms. For one thing, the victim carries the psychological burden alone. She may not seek the help she needs for the repercussions of rape, such as anxiety, depression, suicidal ideation, and PTSD. She deprives herself of the social support that may otherwise circumvent a cycle of self-blame. Health suffers as well. As one college woman in our study commented, "I got pregnant, stopped eating because I didn't know what to do, miscarried, and then decided I deserved to die because of what I had done. An eating disorder and self-mutilation shortly followed" (twenty years old, raped at age thirteen).

Suffering alone can also lead victims to act on their suicidal thoughts. One study of 158 women who had attempted suicide found that an astonishing 50 percent had experienced sexual abuse at some point in their lives. In a one-year follow-up study, those who had experienced sexual abuse were more likely to repeat a suicide attempt than those who had not been sexually abused.[35] Other studies corroborate this link. One found that 33 percent of rape victims had contemplated suicide, compared to only 8 percent of non-victims, and 13 percent of rape victims had actually attempted suicide, compared to only 1 percent of those not sexually victimized.[36] Suffering alone sometimes proves disastrous.

Keeping the rape secret also allows the rapist to escape punishment. It frees a sexual predator to pursue other victims, or, as sometimes happens, to return to his original victim to prey on her again. These ruinous social consequences provide some of the motivation for movements to decrease the shaming of rape victims and urge them to come forward. These are well-intentioned efforts and perhaps they will work in the long run. In the meantime, it is the individual victim who incurs the harms when her rape becomes public knowledge.

Finally, despite their valid motives, when women conceal

their sexual abuse, society, including law enforcement, underestimates its prevalence. One source of criminal justice statistics estimated that only 230 out of every 1,000 rapes are reported to the police, in contrast to 619 out of 1,000 robberies and 627 out of 1,000 assault and battery crimes.[37] The police force is a limited resource. It is deployed not just to catch criminals but also to deter future crimes. Because 75 percent or more of all rapes go unreported, and police resources are allocated in part according to the reported incidence of the crime, fewer resources are devoted to catching rapists than if more rapes were reported. Concealed victimization, while understandable given the suffering caused by its exposure, unintentionally causes insufficient allocation of law enforcement resources, which in turn allows sexual predators freedom to prey on other victims. And because most police officers are men (87 percent in the United States) with a male sexual psychology, they may lack sufficient empathy for rape victims, which leads to greater leniency toward sexual predators.

Many costs of rape are summarized in the words of a victim in one tragic case, whose husband divorced her even though she saved his life: "One night when we were asleep, five armed men stormed our house. They trussed up my husband and demanded that we give them money. Brandishing their guns, they threatened to kill him or rape me if we didn't give them money.... To save my husband's life, I told them to take me instead. All five of them raped me before my husband and children. They looted everything. Since then my husband has abandoned me. There is this culture that when a woman has been defiled, the husband has the right to leave her."[38]

Because sexual coercion is so harmful to women, and has been over the long course of human history, it would defy evolutionary logic if selection had not fashioned defenses in women to prevent its occurrence. Having outlined the evolutionary harms of rape to victims, we now turn to women's defenses to prevent becoming victims and to mitigate costs in the aftermath of victimization.

THE BODYGUARD HYPOTHESIS

Professors Sarah Mesnick and Margo Wilson proposed perhaps the first line of defense, which they termed the *bodyguard hypothesis*.[39] According to this hypothesis, women form heterosexual pair-bonds with men in part to reduce their risk of sexual aggression from other men. Women should be especially attracted to physically large and socially dominant men as mates, according to this view, and this should be especially true in contexts in which they are most at risk of sexual aggression.

Having a formidable bodyguard could function in one of three ways. First, a bodyguard could *deter* men who might otherwise target his mate for sexual coercion. Second, a bodyguard could *intervene* during an attempted act of sexual coercion. Third, a bodyguard could *seek retribution* against the attacker by damaging his reputation, injuring him, ostracizing him, or killing him, sending a strong signal to other men that sexual coercion will have perilous repercussions.

Abundant scientific evidence supports the hypothesis that women do desire physically formidable men as long-term mates.[40] Tall men are consistently seen as more desirable than men average or short in stature. In personals ads, of the women who mention height, 80 percent desire a man who is six feet tall or taller. Ads placed by taller men receive more responses from women than other ads, which explains why men tend to "round up" deceptively when mentioning their height. Tall men tend to have higher status than short men — a finding consistent across cultures. "Big men" in many traditional cultures, meaning men high in status, tend to be physically large.[41] Women are also especially attracted to athletic men with a V-shaped torso, one with broad shoulders relative to hips. This torso shape is a cue to upper-body strength, one of the most sexually dimorphic dimensions of the human body and a strong cue to fighting formidability.

An episode described by anthropologist Napoleon Chagnon captures the bodyguard function of physically formidable men and the consequences of lacking one. While being given a tour of the village by one of the "big men," he passed a hammock in which the brother of the big man was relaxing. As they approached, the tour guide kicked his brother and told Chagnon, "This is the brother I was telling you about whose wife I fucked." The less formidable brother left the hammock and slunk away in shame, not wishing to challenge his more powerful sibling. Women with less powerful bodyguards may be more vulnerable to coercion. Stronger mates are able to protect their partners and deter other men from sexual interloping and aggression.

Women's mate preferences for physically imposing men, of course, do not provide definitive support for the bodyguard hypothesis. Women undoubtedly prefer formidable men for other potential benefits, such as protection for their children, success in defending their coalition from attack, and success in shelter building and large-game hunting. Because big men also tend to rise in status, and people tend to cede resources, such as food, healthcare for children, and prime patches of land, to those at the top of the hierarchy, a woman and her children and kin all benefit. Protection from potentially aggressive other men is only one among several benefits that flow to women choosing formidable mates.

In a somewhat different test of the bodyguard hypothesis, Wilson and Mesnick studied 12,252 women, each interviewed over the phone by trained female interviewers. Questions about sexual assault began with: "Has a male stranger ever forced you or attempted to force you into any sexual activity by threatening you, holding you down, or hurting you in some way?" Subsequent questions dealt with unwanted sexual touching: "[Apart from this incident you have just told me about,] has a male stranger ever touched you against your will in any sexual way, such as unwanted touching, grabbing, kissing, or fondling?"[42] The statistical analyses focused on sexual victimizations that occurred within the twelve months prior to the interview and excluded sexual assaults by husbands or boyfriends.

A total of 410 unmarried and 258 married women reported experiencing one or more of these sexual violations. Marital status had a dramatic effect on sexual victimization. Among women in the youngest age bracket of eighteen to twenty-four, 18 out of 100 single women reported sexual victimization by a stranger, whereas less than half that number — only 7 out of every 100 married women — reported sexual victimization. Although Mesnick and Wilson conclude somewhat sanguinely that their results support the bodyguard hypothesis, they acknowledge that they have not identified the causal mechanism by which married women are less likely to be victims than comparably aged single women. The lower rates among married versus single women might reflect lifestyle differences rather than a bodyguard effect — single women might spend more time in public places such as parties, concerts, or bars rather than at home, making them more vulnerable to rapists. The finding could also reflect individual differences in mating strategies. Single women might be more likely to pursue short-term matings that place them in contexts where they are exposed to greater danger of sexual coercion, and one study found that women who prefer a short-term mating strategy are indeed more vulnerable to sexual victimization.[43]

Romantic partners are not the only bodyguards a woman may have at her disposal. According to Professor Barbara Smuts, women form "special friendships" with men who provide protection. In our studies of opposite-sex friendship, Professor April Bleske-Rechek and I found that women listed protection as a key quality in their choice of male friends.[44] Women also cited physical protection as one key reason for initiating an opposite-sex friendship. And women, more so than men, cited a friend's failure to provide that protection as a key reason for ending the friendship.

Women also cultivate alliances with other women who function as bodyguards. Here is how one woman in the military described this tactic after she heard that a woman had been raped near her own office: "We all figured out where [the other women] lived, and we walked as a group instead of just one or two together."[45] It is far more difficult for a man to sexually coerce a woman whose friend has her back, can fend off an attacker, and can alert other social allies to danger. Moreover, female friends can deter potential predators through the threat of reputational damage. The social costs of being a sexual predator become magnified as word spreads. Men known as rapists have targets on their backs. They risk injury, ostracism, and sometimes murder by enraged allies of the victim.

Parents too can function as bodyguards.[46] Researchers studied the ways in which parents attempted to influence their college-age offspring's sexual and mating conduct. According to both parents and their offspring, parents tended to set earlier curfews for daughters compared to sons. They exerted greater control over their daughters' clothing, restricting their use of skimpy or revealing garb. They more closely monitored their daughters' sexual behavior. And they were more likely to insist on meeting their daughters' prospective dates or boyfriends. Women who live with or close to kin are less vulnerable to sexual coercion than women living apart from or distant from their genetic relatives.[47]

Another source of evidence for the bodyguard hypothesis comes from a line of research that attempted to identify women's rape-avoidance behaviors.[48] Researchers began by asking a large panel of women what actions they performed to avoid being raped. Subsequent statistical analyses on a different sample of women yielded four major rape-avoidance strategies. *Avoid being alone* emerged as one of the most prominent. This strategy included: "I stay around other people when I go out"; "I avoid going to public restrooms alone"; "When I go out, I go with at least one male friend"; and "I avoid walking alone at night." Another study found that 42 percent of women, as contrasted with 8 percent of men, avoided going out alone.[49] A parallel study of college-age individuals in Greece (average age twenty) revealed an even larger gender difference — 54 percent of women, but only 2 percent of men, avoided going out alone.[50]

Women also alert bodyguards when they do go out alone, for example, "I let family or friends know where I am going when I go out." Women take a variety of precautions to decrease their vulnerability when traveling or socializing, and alerting allies is an important one.

The scientific status of the bodyguard hypothesis, of course, requires more direct tests. Are women especially likely to choose large, physically imposing men when living in social circumstances that indicate a relatively higher risk of rape? Are women with formidable mates and friends less likely to be victims of sexual coercion than women with less imposing partners, as the Yanomamo episode suggests? Are women who enlist female allies less likely to become victims? Despite a lack of definitive answers to these key questions, enlisting bodyguards is an important defense against sexual coercion. In an ideal, rape-free world, women would not need bodyguards. In the real world, bodyguards help.

THE WISDOM AND TRAGEDY OF FEAR

Sociologists and criminologists have long observed what they call the "fear of crime paradox." Women are generally more fearful than men of all sorts of crimes — including being physically assaulted and murdered — but are statistically less likely to be victims of these crimes. One obvious exception looms large. Women are far more likely than men to be victims of sexual coercion. One study assessed women's fears of nineteen different crimes.[51] Among women age nineteen to thirty-five, fear of rape exceeded fears of all other crimes by a wide margin. On a scale of 0 (no fear) to 10 (extreme fear), these women gave fear of a rape a rating of 6.81. They expressed less fear of being beaten up by a stranger (4.40), being threatened with a knife or gun (4.34), and having their car stolen (3.49). On the index of *seriousness* of the crime, women rated being raped as equivalent to being murdered.

Some scientists argue that women's fear of sexual assault generalizes psychologically, influencing their overall fear of crime. For example, women fear being mugged or having their house broken into in part because these crimes can be linked with rape — a realistic appraisal given that men who break into houses do sometimes opportunistically rape women who happen to be home, even if their primary goal is robbery. Indeed, when researchers statistically control for fear of rape, the sex differences in fear of nonsexual crimes disappear or even reverse.[52] These findings suggest that fear of sexual assault is a primary driver of women's fears of other crimes, providing one resolution of the fear of crime paradox. Uncovering the logic of why women fear rape so strongly, however, requires a deeper explanation that starts with exploring the nature of fear as an extraordinarily useful emotion rooted deep in evolutionary history.

Humans, like all other mammals, have encountered dangers from which they must escape. Fear is one of the oldest and most conserved mammalian emotions. It is housed in midbrain circuits, specifically in the amygdala, which governs its activation and deployment. Fear evolved to motivate mammals to avoid threats, escape from threats, and combat threats that cannot be avoided. Historical scientific treatments capture this as the "fight-or-flight response." As a species, we have developed specialized fears of specific dangers such as snakes, spiders, heights, and darkness.

These are evolutionarily ancient hazards. Fear kept our ancestors alive. It prevented bone-shattering falls and venomous bites. Without this powerful emotion, many of our ancestors would not have become our ancestors.

Members of our own species also pose special hazards. Historically, none were more threatening than strangers. Stranger fear emerges early in life, at around six months of age, when infants develop the ability to crawl away from caretakers. Infant stranger fear has been documented cross-culturally, including in Guatemala, in Zambia, among Hopi Indians, and among the !Kung San bushmen of the Kalahari Desert.[53] It is a human universal. Infants do not fear all strangers equally. They show special fear of unfamiliar males, suggesting that male strangers have been more dangerous than female strangers. And in infancy, this is statistically true. Just as male lions kill the cubs of their rivals when they displace them, having a strange male in the household, particularly a real or de facto stepfather, is the single greatest risk factor for human infanticide.[54]

Fast-forwarding to the teenage years, strange men pose a special hazard to females—the risk of rape. Stranger fear is highly functional. It commands our attention. Pupils dilate, allowing us to perceive the threat more clearly. Fear makes the heart race and blood pressure rise. Muscles tense. Breathing becomes rapid, sending oxygen to muscles needed for fighting or fleeing.

Rape fear more specifically shows qualities that point strongly to its adaptive value. First, rape is feared by women far more than by men — an obvious observation, but an important one. Few men can deeply understand women's experience of anxiety surrounding the prospect of sexual assault — a psychological disconnect that interferes with men's empathy. As the feminist author Susan Griffin wrote: "I have never been free of the fear of rape. From a very early age, I, like most women, have thought of rape as part of my natural environment — something to be feared and prayed against like fire or lightning."[55] Few men live with this fear except in unusual settings such as prisons.

Women's rape fear is common across cultures, including India, China, and Middle Eastern countries such as Israel.[56] Even in sexually egalitarian cultures like Sweden and the Netherlands, women report experiencing dread of sexual assault.[57] Although cultures differ dramatically in the sexual dangers they pose, in no culture are women entirely free of rape fear.

Age influences the magnitude of anxiety. Fear of sexual assault is more prevalent among young women, precisely those who are most likely to become victimized. A study of women's fears of nineteen crimes revealed the highest fear among the youngest group (age nineteen to thirty-five), with rape fear progressively decreasing with each increase in age bracket. Women in their sixties and up fear home break-in far more than they fear rape, while the reverse is true of younger women. Even at universities, first-year female college students fear rape more than their senior peers, showing that fear roughly tracks victimization probability.[58]

The strong link between youth and probability of sexual assault has potentially important policy implications. Some rape education materials convey the message that all men are potential rapists, that any woman can be raped regardless of age, and that rape is essentially random.[59] The intent behind this message is benevolent — to combat the unfortunately pervasive tendency to blame victims of sexual assault.[60] Exposure to these messages, however, may have the unintended consequence of making some women feel helpless to defend themselves, leading simultaneously to a higher fear of rape and fewer defensive precautions.[61] The fact that women's rape fears track their age-linked vulnerability suggests that their anti-rape psychological wisdom may be a better guide than factually incorrect educational materials, however well intended those may be.

Women who perceive themselves to be physically strong fear

stranger rape less than women who perceive themselves to be weaker.[62] Women who perceive that they would be able to successfully escape from an attempted rape show less rape fear than women who perceive that their escape attempts would be futile.[63] Physical formidability, enhanced in the modern era through exercise and classes in self-defense, appears to reduce women's trepidation about sexual assault.

Women fear rape at night more than during the day, accurately tracking when rapes are most likely to happen.[64] Night also accompanies other rape risk factors, such as attending parties, drinking alcohol, and encountering unknown males. And as it does for other crimes, darkness affords criminals more opportunity to surprise victims and escape after committing their crimes without capture.

Women who personally know other women who have been raped report a higher fear of being raped themselves.[65] This finding reveals a psychology of fear that gets adaptively ramped up by social knowledge of victims in one's social circle. Here is what one woman in the military reported: "After we learned that we were going to deploy, one of my cousins told me that she had been raped when she had deployed. And then other women in my barracks started talking about women they'd known from basic training who had been raped in Afghanistan. I started getting really scared."[66]

Personal knowledge of rape victims could indicate a higher prevalence of male rapists in the vicinity, a higher probability of personally becoming a victim because female friends tend to share similar risk factors such as age and social circles, or emotional contagion activated by empathy for the victim. Women who have been sexually harassed, such as being leered at, propositioned, followed, or hassled by men in hotels, also reported higher levels of rape fear.[67] "Sexual harassment," to quote one researcher, "may reinforce the idea for women that they are potentially the sexual prey of men."[68] Women's level of rape fear, in short, appears to be psychologically calibrated both to personal knowledge of other women who have been raped and to their personal experiences with being sexually harassed.

Evidence for adaptive calibration also comes from a strong link between women's rape fear and their reports of behavioral precautions they take to avoid rape. Women who are especially fearful of rape are more likely to avoid being alone with men they don't know well, decline rides from men, leave when a man comes on too strong sexually, avoid outdoor activities when alone, and exercise more caution in their alcohol consumption. Women living in neighborhoods with a high incidence of rape report more fear than those living in safer neighborhoods. A study of Dutch women found that women living in large urban areas such as Amsterdam experience more rape fear than women living in small towns or suburbs. Rape rates are indeed higher in metropolitan than non-metropolitan areas.[69] Women's fears, in short, track realistic dangers, some modern and some evolutionarily ancient.

Perhaps the most striking evidence of the protective function of fear comes from what is called the "second crime paradox," which is that women are more fearful of being raped by strangers than by acquaintances. One study found that college women estimated that 43 percent would be raped by strangers, and this is surely an overestimate. Even the highest estimates put women's chances of attempted or completed rape at 25 percent, and of these, an estimated 10 percent will be committed by strangers.[70] The actual percentage of college women experiencing rape or attempted rape *by strangers* is estimated to be closer to 2.5 percent, rendering women's estimates of 43 percent off base. Stranger rape is relatively rare, as noted earlier, accounting for only 10–20 percent of all rapes, compared with acquaintance rape, which is far more common, roughly 80–90 percent. Some social scientists conclude that women's fears are miscalibrated to the realities of rape, but I think that conclusion may be too hasty and needs context to be interpreted correctly.[71]

Before we conclude that women's stranger rape fears are maladaptive, we must consider an alternative interpretation — that women's apprehensions about strangers are actually highly effective. Women's stranger fear motivates precautionary behaviors, dramatically reducing the rates of stranger rape below what they would be without these functional fears.

Even if women are wrong ninety-nine times out of a hundred about a stranger's rape intent, the one time they are right aids in avoiding rape and its calamitous costs. According to this view, women's stranger fears actually work reasonably well, preventing some sexual assaults, even at the expense of the energy expended to avoid strange men who may be entirely harmless. Women's rape fears may be biased in the statistical sense of overestimating stranger rape probability, but they are adaptively biased. They would have worked well in ancestral environments in which many rapists were likely to be strangers. They also may be functioning well in the modern environment, lowering the current rates of rape by unknown men. Given that a third of all men indicate some non-zero likelihood that they would rape if they could be assured that there would be no consequences, the actual rates of stranger rape might be considerably above 2.5 percent if women did not have these adaptive fears. Moreover, women appear to be reasonably accurate in their estimate that 37 percent of men would rape "if they thought they could get away with it," which roughly coincides with men's self-reports of same.[72]

Another perception-reality disconnect, at least at surface glance, is that many women conjure vivid images of serious injury, or even death, accompanying rape.[73] In reality, women rarely experience serious physical injury, and most rapists use just enough force to accomplish the crime. As noted in Chapter 7, rapists rarely murder their victims. Before concluding that women's fears of physical injury are irrational, we must again add nuance and context. First, stranger rapists frequently issue *threats* of serious injury, and sometimes threats of death, unless their victims comply.[74] Second, women's fear of collateral injury may add urgent motivational fuel to avoid putting themselves in danger, further lowering the odds of falling victim. Support for this notion comes from a study that found that a significant predictor of an elevated fear of rape is a woman's belief that collateral physical injury often accompanies rape.[75]

In short, what seems at first glance like irrational fears may instead be hallmarks of a well-honed adaptive defense when it comes to the harms of stranger rape and physical injury accompanying rape. But although women's rape fears appear to be adaptive, lowering their odds of becoming victims, these fears come with heavy costs. As feminist scholars have noted, fear keeps women indoors, especially at night, and prevents them from the freedom of movement that men enjoy. The attention and effort women allocate to anti-rape vigilance detracts from the attention and effort they could allocate to other challenges of life, including taking night classes, socializing freely, and advancing in their chosen profession. Women's rape fears have additional social costs when women feel forced to prioritize friends and mates based on their ability to offer bodyguard protection, rather than on other important qualities.

Gavin de Becker's book *The Gift of Fear* notes that fear reflects ancestral wisdom; without it, our ability to survive and thrive undoubtedly would be jeopardized.[76] Given the costs women incur from rape fear, however, eliminating the actual threat of rape would be a far better solution. Now, armed with more accurate scientific knowledge about the damage it causes, the men most likely to inflict these costs, and the conditions that prevent men from doing so, this solution is approaching the realm of possibility. Even so, women may remain tethered to some of the fears of their ancestors. Our psychology of fear has been valuable in the past and continues to help women avoid at least some sexual traumas today. It is a tragedy of profound proportions, however, that women are forced to experience anxiety and dread about the sexual predators in their midst.

FROM VIGILANCE TO TONIC IMMOBILITY

Women's rape fears do not merely reside in their interior psychological experiences. Fear motivates a suite of tactical actions, starting with vigilance. Assessing threats and risks is a key function of vigilance. Like fear, vigilance is evolutionarily ancient. It is especially important in predator-prey arms races such as that between cheetahs and gazelles. In every generation, the prey who are less attentive to danger get picked off while the more alert and observant escape predation and pass on keener vigilance to their offspring.

Vigilance is an adaptation heavily attuned to risk and uncertainty. Risk detection prompts physiological arousal. Arousal prepares the body for rapid, high-exertion defensive action by releasing adrenaline, which makes the heart race, increases blood circulation, accelerates breathing, and metabolizes carbohydrates for use by the muscles. It prepares the body to act if an uncertain risk turns to real danger.

Threat activates alertness and watchfulness. The visual sense becomes galvanized as individuals scan their environment to pinpoint the threat. Watchfulness engages more than vision; it enables the detection of acoustic threat cues. The *auditory looming bias* causes people to overestimate the closeness of fast-approaching sounds, which are more dangerous than quickly receding sounds — this aids in detection, preparation, and threat avoidance. Even the sense of smell becomes more sensitive, since olfactory cues can convey information about the size and formidability of a predator as well as its proximity. Visual, acoustic, and olfactory cues to threat are all more easily detected in a state of *attentive immobility*, colloquially captured by the term "freezing." Stilling the body heightens the senses. It provides an opportunity to assess the potential dangers in the environment — where are they coming from, how close are they, what are the routes for escape.

Vigilance, in short, mobilizes the mind, ramps up alertness, and activates scanning for the source, magnitude, and proximity of danger.[77] Threat detection leads to evaluation of defensive strategies. Threats often come from other people; for women, the risk of rape comes mainly from men. The array of defensive actions in women's arsenal is shown in the following table.[78]

WOMEN'S DEFENSES AGAINST SEXUAL COERCION

Avoiding Assault

Defense	Adaptive Goals	Typical Behaviors
Secure Bodyguards	Deter sexual coercers	Maintain proximity to kin, allies, or formidable mates
Rape Fear	Avoid dangerous situations Avoid dangerous men	Vigilance Avoidance

Escaping During Assault

Defense	Adaptive Goals	Typical Behaviors
Attentive Immobility	Detect source of danger Detect magnitude of danger Calculate means of escape	Freezing Hyperalertness

Defense	Adaptive Goals	Typical Behaviors
Withdrawal	Escape danger	Fleeing
Refuging	Hide from danger	Sheltering in safe harbor
Alarm Calling	Alert bodyguards	Screaming
Aggressive Defense	Ward off attack	Fighting
Appeasement, Pleading, Acquiescence	Deter aggression Minimize harm from aggression	Affiliating Submitting
Tonic Immobility	Minimize harm from attack when escape is not possible	Body paralysis

Reducing Harm After Assault

Defense	Adaptive Goals	Typical Behaviors
Aftermath Concealment	Avoid blame for assault Prevent damage to reputation Avoid relationship damage	Camouflaging bruises Suppressing information Washing/douching
Post-Trauma Stress	Avoid circumstances in which sexual coercion occurred	Intrusive memories Flashbacks Withdrawal Sheltering

DEFENDING AGAINST SEXUAL COERCION

Avoiding sexually risky situations to begin with is often women's first line of defense. Many women *routinely* engage in risk-avoidance maneuvers that remove them from harm's way. In one study of urban women, 41 percent reported "isolation tactics," such as not going out on the street at night. A full 71 percent reported using "street-savvy tactics," such as wearing shoes that enable running away should they be pursued. A study in Seattle found that 67 percent of women avoided certain dangerous locations of the city; 42 percent reported not going out unaccompanied; and 27 percent sometimes refused to answer a knock on their door. A study of Greek women reported that 71 percent avoided venturing out alone at night, and 78 percent shunned dangerous locations within the city.

Women also show wariness around men who talk about sex a lot, men who appear to be sexually aggressive, and men who have a reputation for sleeping with a lot of women. When dating men they don't know well, women report choosing public places for their rendezvous and informing their friends or family about their plans should they lead to danger. They intentionally avoid giving mixed sexual signals to certain men. They sometimes carry Mace, whistles, or weapons. And they sometimes limit their drinking around men they don't know well. The strategy of avoidance, however, comes with its own costs. Women are forced to forgo freedoms, opportunities to acquire resources, and even opportunities for beneficial mating.

Withdrawal or escape when confronted with a threat or potential threat is a second line of defense. In a study of women's rape-murder fears, many reported evasive action when a strange man appeared to be staring at them or following them in unsafe areas.[79] They crossed to the other side of the street. They fled on foot, walking briskly or breaking into a run. They jumped into their cars or hastily entered their apartments and quickly locked the doors. Women sometimes sought the protective cover of other people, such as speed-dialing a friend or quickly entering a crowded movie theater. These behaviors might feel familiar to many women reading this chapter. When faced with uncertainty about the reality or magnitude of threat, it's often better to be safe than sorry.

After successfully escaping a rape threat, women often engage in *refuging*. Threatened women seek safe spaces such as their homes or workplaces, lock doors to prevent a breach, or seek solace with allies who function as bodyguards. A couple's home, as well as the homes of boyfriends, girlfriends, parents, and coworkers, can provide a protective haven against harm.

When escaping or refuging is not an option, *aggressive defense or fighting* becomes key. Studies of victims of attempted sexual assault find that rapid high-intensity fighting often prevents a rape. One study of 150 sexual assault victims age sixteen and older found that, among women using forceful resistance such as biting, struggling, punching, kicking, or using a weapon, 55.5 percent escaped being raped compared to only 6.5 percent of the women who did not resist at all.[80] Fleeing or running away, in this study, proved the second most-effective strategy, with 45 percent who used this tactic escaping rape. Women in this study who forcefully resisted were not more likely to be physically injured than women who did not resist. Although causality cannot be inferred, the link between forceful physical resistance and rape avoidance has been replicated in many studies — screaming loudly and yelling for help, forms of alarm calling, while aggressively resisting appears to be especially effective.[81]

Although many of these studies examined victims of stranger rapes that were reported to the police, other studies suggest that forceful physical resistance is also effective in thwarting rapes with known assailants such as acquaintances or boyfriends.[82] Moreover, the confidence with which women carry themselves and the determination with which they stride can deter some rapists. Interviews with incarcerated rapists reveal that many avoided selecting victims they perceived to be "fighters." One rapist made this observation: "[If] she walks on her heels and not her toes, I know she's not fit and probably wouldn't run or fight."[83] Another reported, "If she's not watching what's happening all around her, then [she] doesn't know how to handle herself, how to use things to hurt me or get me caught."[84] They also reported lower completion rates with victims who actively and forcefully resisted.[85]

Although the evidence is mixed, some studies suggest that taking self-defense classes and teaching women effective resistance tactics boosts their confidence to resist and may lower their risk of rape.[86] Scientists who study rape-avoidance effectiveness conclude that some tactics, such as offering mild verbal resistance or no resistance at all, often prove ineffective. Weak or low-resistance tactics seem less likely to deter men who attempt to rape.

Appeasement is another defensive strategy when escape or active resistance is not possible or is unlikely to be effective. A woman can plead with her attacker and try to talk him out of the attack. Some women display submissive postures or behaviors in an effort to indicate that they pose no threat to the attacker. Unfortunately, nonforceful verbal strategies like pleading and appeasement are not typically successful in deterring most rapists. Alice Sebold, author of the poignant and disturbing rape memoir *Lucky*, tried to appease her attacker by repeatedly pleading, truthfully, that she was a virgin. It did not work. He brutally raped her despite her entreaties. More systematic studies are needed, however, on the relative effectiveness of these defensive tactics.

When escape or defense is not possible, as when women are tied up, shackled, trapped, physically overpowered, or threatened with a weapon, the last resort is *tonic immobility*. Physiologically, tonic immobility is linked with a drop in blood pressure, numbing throughout the body, and analgesia or insensitivity to pain, likely caused by the release of natural opioids in the bloodstream.[87] It is accompanied by catatonic-like body posture, tremors, periodic eye closure, feeling motorically paralyzed, and being unable to cry out for help or resist an attack.

Tonic immobility is a common response to high-intensity fear-inducing events from which escape is not possible. It is similar to freezing but differs in that the body experiences paralysis. Whereas freezing heightens awareness about the source and nature of danger and opportunities for escape, tonic immobility occurs when there is little or no chance of fleeing or winning a confrontation. Feelings of entrapment can be exacerbated by elements such as the presence of a weapon wielded by the attacker, size and strength disparities, verbal threats of violence, threats of death if escape is attempted, and characteristics of the physical environment such as the lack of visible escape routes.[88] The combination of entrapment and extreme fear appears to produce the highest probability of tonic paralysis.

Tonic immobility is so common in rape that it has generated the phrase "rape-induced paralysis." One study found that 37 percent of rape victims experienced tonic immobility during the attack. Other studies estimate that about a third of all rape victims experience some degree of tonic immobility, although some put it as high as 41 percent.[89] A rape victim described her experience in this way: "I felt faint, trembling, and cold.... I went limp."[90] Another victim said, "My body went absolutely stiff."[91] Tonic immobility is rarely the first line of defense. Rather, it comes online when other defenses, such as appeasing the assailant, screaming to alert others, fleeing to escape, or fighting to deter the attack, have not afforded success or are not possible because the victim feels extreme fright combined with physical entrapment.

From an evolutionary perspective, some authors hypothesize that tonic immobility is an adaptation to minimize attack ferocity when escape is not possible. Some prey animals, for example, "play dead" when attacked. Since predators are often sensitive to the motion of potential prey, when prey do play dead, the predators sometimes loosen their grip on the victims, opening up the opportunity for escape. Analogies between prey animals and rape victims in tonic immobility, like all analogies, have limited utility. Predator-prey interactions involve different species; rapist-victim interactions involve the same species. Predators seek prey for food; rapists do not. Moreover, whether tonic immobility is an adaptive or maladaptive response to an attempted rape has not yet been determined. Does immobility lower the odds of rape for women, given that some rapists become aroused by victims who struggle? Does immobility lower the odds of physical injury? The answers remain unknown. What is well established, however, is that women's tonic immobility is sufficiently common that scientific analysis of its potential adaptiveness is warranted.

It is imperative to bear in mind that tonic immobility is an *involuntary* response that appears to be automatic and unlearned.[92] This is especially important because the police, judges, and juries often question whether a reported rape was consensual sex if the victim did not ferociously or violently resist the attacker. In my view, it is tragic that victims who lack physical cuts and bruises encounter skepticism of their reports of rape. They receive less social and emotional support. Women who experienced tonic immobility are also more likely than other rape victims to blame themselves, to experience guilt, and to feel like they could have done more to prevent the attack.[93] Disseminating the knowledge that tonic immobility is a frequent response to rape, that it occurs most often when victims experience extreme fear and entrapment from which escape is not possible, and that it is an involuntary response may help police, judges, juries, friends, parents, and acquaintances to stop blaming rape victims. Ideally, this knowledge would also prevent victims from blaming themselves.

Another source of victim self-blame occurs when a sexual assault victim experiences vaginal lubrication during the assault — a physiological response that occurs in an unknown number of rape victims. Some have speculated that vaginal lubrication is an adaptation to prevent injury, scientifically labeled the *preparation hypothesis*.[94] Vaginal penetration without lubrication can cause vaginal bruising and tearing, especially during nonconsensual sex. These injuries can damage a woman's reproductive tract and increase the odds of infections. Infections can cause pelvic inflammatory disease, which is created by sexually transmitted bacteria that infect the fallopian tubes or ovaries. It sometimes produces infertility. Lack of vaginal lubrication during penetration, in short, can be extremely dangerous to a woman's health, her reproductive tract, and her ability to bear children. An experiment to test the preparation hypothesis had women listen to recordings of consensual and nonconsensual sex.[95] Researchers recorded genital response and subjective reports of sexual arousal. Women experienced vaginal lubrication in both conditions, even though they reported finding the nonconsensual episodes to be extremely unpleasant, anxiety-provoking, and not at all subjectively sexually arousing.

Women's increased vaginal blood flow and lubrication, like tonic immobility, are automatic responses. They do not indicate consent. They do not indicate that the woman finds it enjoyable. They do not indicate that the woman is turned on or sexually aroused. But victims who experience them often feel embarrassed about reporting them, and for good reason. Blame from others and self-blame may follow. Self-blame, poorer-quality social support, and tonic immobility during a sexual assault appear to be linked to post-traumatic stress disorder (PTSD) and depression in the aftermath.[96]

CONCEALMENT IN THE AFTERMATH OF SEXUAL ASSAULT

Of all the violent crimes, rape is probably the most underreported.[97] One survey study found that more than 50 percent of all victims never revealed the sexual assault to anyone, not even to their closest friends, within five years of the crime.[98] At least 75 percent of all rapes are not reported to police, and some estimate the underreport rate to be as high as 90 percent. Report rates are lower for acquaintance rapes than for stranger rapes. The reasons for underreporting are many and complex. They include self-blame and blame by others. They include suspicions that women lie about rape, so victims worry that they might not be believed. They include the grueling process of grinding through the legal system, being forced to relive the trauma in depositions and court testimony, being subjected to skeptical cross-examination by defense lawyers, having their personal lives exposed, and anticipating a low likelihood that the accused will be convicted. Even if convicted, rapists often do little jail time.

Chanel Miller, the victim of Brock Turner, the Stanford athlete convicted of sexually assaulting her while she lay unconscious behind a dumpster, went through grueling medical and forensic examinations, four years of psychological anguish, a shattered romantic relationship, alienation from friends, employment disruption, repeated police interrogations, repeated defense lawyer depositions, and humiliating court testimony about the minutiae of her personal life — all to see the perpetrator receive a slap on the wrist.[99] Brock Turner received a mere six-month jail sentence, and he was released after serving just three months. He went from being a promising Olympic-level swimmer at Stanford to working at a factory job earning just over minimum wage and must register as a Tier 3 sex offender (the most severe tier) for the rest of his life.

On top of the costs victims face when a sexual assault is revealed, disclosure of the crime often leads to victim blaming. And it's not just police. Friends, family, and romantic partners often blame the victim as well. They question why the woman dressed in a certain manner, why she was at a party, why she drank alcohol, or why she flirted with a man she did not know well. Is it any wonder that rape victims are reluctant to report or reveal?

For all these reasons, *concealing* a rape is one coping strategy women use to avoid incurring these additional harms. Revealing a rape can exacerbate a victim's feelings of shame and humiliation. Rape victims cannot be blamed for wanting to avoid being retraumatized by the repercussions. The urge to conceal may explain why victims often feel a pressing desire to shower, bathe, scrub, douche, and cover their bruises, despite well-intentioned police who encourage just the opposite to preserve the forensic evidence. These actions conceal the crime rather than reveal it.

If the rapist resides in his victim's geographical or social milieu, the victim risks running into the perpetrator and being psychologically retraumatized. Perpetrators sometimes feel free to revictimize the original target, to victimize other women, or to grow even bolder and more audacious in perpetrating sexual assaults. The costs of concealment, in short, are incurred both by the original victims and by future victims.

POST-TRAUMATIC STRESS SYMPTOMS: DISORDER OR ADAPTATION?

Experiencing a sexual assault leads to post-traumatic stress in many victims.[100] Common symptoms include flashbacks to, and vivid imagery of, the traumatic experience; sleep disturbances, such as insomnia; avoidance of cues linked to the assault, such as places where it occurred or people who trigger a reminder; reduced activity driven by fear of venturing outside; difficulty concentrating on work or school; nightmares about the assault; and hypervigilance to cues of potential danger. This cluster of symptoms is common in both rape victims and war combat veterans. One study of ninety-five rape victims, for example, found that 94 percent of them met the diagnostic criteria for PTSD on initial assessment after the assault; 65 percent on a four-week follow-up assessment; and 47 percent after three months.[101] PTSD typically has been labeled a disorder, especially if it persists for more than a month.

An evolutionary perspective, however, raises the question of whether PTSD is truly a maladaptive disorder in an evolutionary sense. Might it instead be part of an evolved defense for coping with the aftermath of sexual assault and other traumatic events? Professor Chris Cantor analyzed the components of PTSD for their functional value, drawing on animal research on anti-predator defenses.[102] He first considers the symptom of *avoidance*, shunning certain places or social situations. These might be adaptive changes in behavior that function to avoid encountering the original threat or similar sources of threat.

Another adaptive response common in post-traumatic stress, Cantor suggests, is *refuging*. Remaining within the safe shelter of one's home is a way of avoiding not just threat reminders, but actual threats themselves. Although this is considered a symptom of a disorder from a psychiatrist's perspective, it may be an adaptive defense that lowers the odds of encountering an uncaught sexual predator. Flashbacks could be your brain reminding you, "Don't go over that hill."

Tonic immobility *during* the assault is one of the key predictors of having PTSD symptoms *after* the assault.[103] There could be several reasons for this link. One is that tonic immobility during an assault is typically accompanied by extreme levels of fear, which may carry over to the post-assault psychology of the victim. Another centers on the other important trigger of tonic immobility — the sense of entrapment. Feelings of entrapment, like heightened fear, may persist after the assault, leading to avoidance of any situations linked with the assault, however probabilistic. Even if the risk of another assault is objectively low, the harms of encountering the horror of a repeat episode may be so severe that it pays adaptive dividends to err on the side of extreme caution.

Women who perceive that they are not receiving social and emotional support following the attack are more susceptible to the symptoms of PTSD.[104] Perhaps the lack of social support provides a cue to the victim that she lacks protective bodyguards, which renders future revictimization more likely. The amplified fears and hypervigilance following a sexual assault among victims lacking support may be an adaptive response to a realistic appraisal of future threat. Activating a woman's social circle to ramp up emotional support may help to alleviate symptoms of PTSD by actually lowering the woman's perceived or real future threat.

PTSD, in short, might not be a dysfunction from an evolutionary perspective. It may be an adaptive defense that helps to create distance from real threats and cues that signal real threats. Such a common psychological response to sexual assault calls into question the frequently used labels of "disorder," "dysregulation," and "inappropriate fear reaction."[105] The possibility that fear-induced hypervigilance, avoidance, and refuging may actually help victims avoid encountering future threats suggests that the adaptive defense hypothesis proposed by Cantor should be seriously considered. Past threats are good predictors of future threats, so it may make supremely good sense to be fearful of cues portending ongoing threats.

If the adaptive defense hypothesis turns out to be correct, treatment implications may follow. Treating the symptoms of PTSD, rather than dealing with its sources, may be like turning off a sensitive smoke alarm rather than reducing the prospect of a future fire. One path forward is an educational one. Rather than being told that "something is wrong" with them, as often occurs and which can further a cycle of self-blame, some victims may feel relieved to know that symptoms of PTSD are a normal adaptive response that helped ancestral women to avoid future dangers. Alleviating victim suffering, of course, is important. The highest priority should be given to avoiding the source of that suffering — reducing the rape rate to zero.

CHAPTER 9

MINDING THE SEX GAP

> Treating different things the same can generate as much inequality as treating the same things differently.
> — Kimberlé Crenshaw

EVOLVED SEXUAL ADAPTATIONS ARE NOT behavioral inevitabilities, which is one source of worry that some have about tracing the origins of sexual conflict to our evolved psychology. The concern is based on a misunderstanding that erroneously conflates "evolved" with "inevitable." One important reason that men's adaptations are not invariantly expressed is that men have multiple components to their sexual psychology. Men have the capacity for deception, but also the capacity for honest courtship. Men have inclinations for low-cost opportunistic sex, but also a deep psychology of love, attachment, and long-term committed mating. Knowledge of the multiplicity of these mating components gives us the power to activate some while keeping others quiescent. One key to keeping the crueler components dormant depends on identifying the personal, social, and economic circumstances that activate the good ones and inhibit the expression of the bad ones.

This chapter pans back to examine broader personal and cultural issues wrought by sexual conflicts and suggests ways for reducing them in the modern world. I put forward my tentative suggestions as worthy of further consideration and future evidence-based testing, not as definitive solutions. I am convinced that deep knowledge of our evolved sexual psychology, and especially the ways in which that psychology differs on average between women and men, is indispensable for reducing conflicts between the sexes that have been ongoing for millions of years. These conflicts are evolutionarily ancient, but perhaps for the first time in history we can develop tools to reduce them.

One sign of progress comes from the fact that victims of sexual violence are coming forward and refusing to be silenced. A prominent example occurred when the many actresses victimized by the serial sexual predator Harvey Weinstein made his crimes known. He has been convicted of multiple counts of sexual assault that will likely imprison him for the rest of his life. Another example is illustrated by vocal alleged victims of Ghislaine Maxwell, who has been charged with procuring underage girls for the sexual predator Jeffrey Epstein. Maxwell strongly denied all charges, and at the time of this writing, she is behind bars awaiting trial. This highlights the often overlooked fact that women sometimes ally themselves with men in perpetuating sexual abuses.[1] Another sign of progress is that many institutions are imposing a zero-tolerance policy for sexual harassment in the workplace.

Sexually antagonistic arms races, in which members of each sex have evolved to influence members of the other to be closer to each person's individual optimum, are now enacted in rapidly changing cultural worlds. Internet dating apps open up an explosive array of options for choosing an ideal mate. At the same time, they create mating-option paralysis and provide novel venues for sexual deceivers and predators. Online pornography allows individuals to explore creative sexual possibilities they never knew existed but also creates entirely unrealistic expectations for real-life sexual interactions. Sex dolls, sex robots, and virtual sex technology open up innovative possibilities for sexual gratification and may diminish some of the rage of incels or even reduce rape rates but also may exacerbate the harmful sexual objectification of women.

Sexual conflicts take psychological tolls. They drain emotional energy, create intrusive thoughts, waste cognitive effort, intensify anxiety, magnify sexual regret, fuel rumination, and squander psychological resources needed for grappling with other problems of living. Women especially suffer, as we have seen, from depression, blows to self-esteem, lowered perceived mate value, eating disorders, PTSD, relationship dissolution, professional impairment, compromised sexual functioning, suicidal ideation, and occasionally death. Any reductions in sexual conflict we can achieve will diminish these horrific damages. The first step involves a deep understanding of sex differences in our sexual psychology.

MINDING THE GAP IN SEXUAL PERCEPTIONS AND MATING EMOTIONS

One foundation for bridging the chasm is recognizing that the gulf or gap between men's and women's sexual psychology is real. We have explored many psychological sex differences throughout this book, such as in the desire for sexual variety, inclinations for impersonal sex, the misperception of sexual interest, upset about different forms of sexual deception, and the use of violence as a means to sexual ends. Although women and men share many elements of their mating psychology, including the capacity for love, attachment, and commitment, failure to recognize the differences is a key impediment to reducing conflict.

The differences reveal themselves most starkly in the realm of sexual perceptions and sexual emotions. Men's misperception of women's sexual interest based on a woman's mere friendliness — a smile, eye contact, or an incidental touch on the arm — ushers in anger on both sides. It angers men who feel "led on." Women seem to be giving them clear signals of interest. Men act on them. And when those actions are spurned, men feel embarrassed, humiliated, and resentful.

From a woman's perspective, a man's misperception that she might be sexually interested creates costly problems that offer no perfect solutions. She sometimes feels forced to disabuse him of his mistaken inference, yet with social skill that evokes the least retribution. She attempts to minimize a man's fury at being spurned. Deflection tactics such as telling him that she has a boyfriend or that she doesn't date men in her workplace help women walk that fine line. But they risk maintaining the burning ember of a man's hope that she would be interested if she were not in a relationship or they were not co-workers.

Men need to understand the sexual over-perception bias that afflicts their mating minds. They should know that, most of the time, the women smiling at them are merely being friendly or polite, not signaling sexual interest. At the same time, women need to know that many men are inclined to err in inferring sexual attraction based on the most minimal ambiguous cues. And they should know about their own sexual under-perception bias — that men may be more sexually interested in them than they fully realize. Educating both sexes about these predictable sexual misperceptions can be a starting point. You cannot use introspection about your own mating mind to accurately infer the mind of the other. Self-analysis of our own psychology is often a poor guide, especially when it comes to sexual attraction.

Education about sex differences in the mating emotions comes next. Consider the emotion of *sexual disgust* — the things that repulse you from a sexual perspective. Are men aware that women find more things sexually disgusting than they do?[2] Judging from the number of men who send unsolicited pornographic images of their decontextualized genitals to women, the answer is a resounding "no." Slightly more than half of millennial-age women in the

United Kingdom have received such images, and 78 percent of these were unsolicited. Roughly 27 percent of millennial-age men in the United Kingdom admit to having sent them.[3] Women's most common reaction to receiving them is "gross," an adjective used by 49 percent of millennial women. On the flip side, 30 percent of men think that women will find these images "sexy," but only 17 percent of women use that adjective. Stated differently, 83 percent of women do not find these images at all sexy, and half find them repulsive. In short, some men commit a major mind-reading error. They fail to understand that many women are sexually disgusted by photos of context-free male genitals. Receiving these unsolicited images is a form of sexual contact. Like other forms of sexual contact, consent from the receiver should be in order.

Sending unwanted sexual images is a modern form of sexual harassment. These sometimes come from total strangers. Some women subway riders in New York City report receiving unwanted dick pics through AirDrop.[4] Because AirDrop has a preview feature, receivers are forced to view the images before making the decision about whether or not to receive them. Some lawmakers are taking action to make sexting without consent illegal. In 2019, Texas became the first state to ban sending sexually explicit images without the consent of the receiver.[5] It's a misdemeanor punishable by a fine of up to $500. Reducing this form of sexual conflict through legal means is one strategy. Another is educational — enlightening men about the fact that, although they may find context-free photos of women's body parts to be sexually arousing, most women find photos of men's genitals irritating or revolting. Using one's own sexual mind as an infallible guide for inferring the sexual minds of others is an error. In this case the error evokes sexual disgust in women, often precisely the opposite of the sender's intention, which is usually to induce sexual interest in the woman or to receive nude selfies from her.

Sexual disgust is only one mating emotion that shows stark gender differences. As we saw in previous chapters, women on average exceed men in anxiety about body image, the strength of the link between love and sex, feelings of hurt and upset about casual hook-ups, and the anguish of regret over mistaken sexual choices. Men on average exceed women in the desire for sexual variety, the willingness to have impersonal sex, the overpowering effect of visual stimuli to provoke lust, and regret over missed sexual opportunities.

Another disconnect between male and female sexual psychology centers on sexual coercion. Scientific studies show convincingly that men vastly underestimate the emotional horrors that raped women experience. Real-life examples illustrate men's underestimation. The father of Stanford swimmer Brock Turner, convicted of sexual assault, pleaded with the judge for probation rather than prison for his son, commenting that his son had already paid a steep price for "20 minutes of action out of his 20 plus years of life."[6] The cost to Chanel Miller, Turner's unconscious victim, was more than a mere twenty minutes. She spent years sitting through depositions and testimony and experiencing character-impugning questions about her sex life from defense lawyers, not to mention the end of a romantic relationship and public humiliation. She may carry lifelong psychological scars that most men cannot fathom.

Educating men about the harms women suffer from sexual assault, from anxiety to PTSD, should be required. One possible tool in this educational agenda involves getting men to imagine that their sisters, mothers, partners, and female friends become victims. While this mindset may ultimately prove problematic if taken in isolation, it could serve as a stepping-stone for some men who have trouble grasping the fact that a woman should elicit empathy as a human being, not just as defined by her relationship to a man.

Perhaps the rage men experience, or can imagine experiencing, when one of their loved ones is sexually assaulted can be co-opted in the educational effort to help men understand. The fear of rape that many women experience and the emotional trauma rape victims suffer during and after an assault have *no real parallels in men's minds.* Some might object that men should have similar outrage over all victims, not just loved ones. That is a worthy goal. In my view, if progress can be made toward that goal by leveraging men's deep empathy for their daughters, sisters, and female friends, then that's a start toward broadening the circle of empathy for all women.

Having studied sexual conflict for many years before writing this book, I thought I understood the psychological damage rape victims experience. After all, I had close female friends who privately trusted me and disclosed their abusive experiences. I fancied myself as unusually empathic. Toggling between the scientific studies that reveal PTSD and depression and the moving memoirs of rape victims such as Alice Sebold (*Lucky*), Chanel Miller (*Know My Name*), and Helena Valero (*Yanoáma*), the raped women of the Republic of Congo, and the heart-wrenching descriptions of forty-nine rape victims in a study conducted in my own lab, I noticed a profound shift in myself. I discovered that I too had vastly underestimated the psychological toll that sexual assault inflicts on women. Undoubtedly I still cannot fully grasp, after decades of study, the psychological toll it takes on women. Men may think they understand, but I don't think they ever can fully. I hope that this book helps men move a little closer to bridging the gap.

Subway stations, or Tube stations as they are called in Great Britain, have posted signs: *Mind the Gap.* These help people to successfully traverse the fissure between platform and subway car. Successful cross-sex mind reading requires minding the gaps in sexual perceptions, sexual inferences, and sexual emotions.

CLOSING THE SEX GAP THROUGH LAWS AND POLICIES ON SEX CRIMES

Stalking laws and policies typically hinge on the psychological state of the victim and a "reasonable person" standard. As an example, here is how the University of Texas defines stalking: "A course of conduct directed at a specific person that would cause a reasonable person to fear for the person's own safety or the safety of others or would cause that person to suffer substantial emotional distress."[7] A critical problem arises when a reasonable woman differs from a reasonable man.

We know from scientific research on stalking that women perceive patterns of persistent unwanted romantic pursuit as more fear inducing and emotionally distressing than do men. We also know that women typically experience greater fear than do men when stalked by a former romantic partner after a breakup. So how should the laws invoking the "reasonable person" standard be written? Should two different laws be created, one for reasonable women and one for reasonable men? Should the laws split the difference and average the two?

There has often been a push to write laws in the most sex-neutral manner possible, which is certainly a laudable goal. In this case, however, a sex-neutral law can harm women, especially if the judge adjudicating the case is a "reasonable man" and consults his own intuitions about fears and emotional distress, or if the jury is composed of a combination of "reasonable men" and "reasonable women." By analogy, the field of medicine belatedly discovered that the same dosage of lorazepam (a common anti-anxiety drug) has a far more powerful and dangerous effect on women than on men, even correcting for body weight. Failure to consider the differences between men and women can harm women if they are more sensitive to the drug, and studies showed that women indeed have been given inappropriately high doses.[8]

The legal harms to women of using a "reasonable person" standard are particularly profound when it comes to sexual crimes such as sexual harassment, stalking, and rape. We know that men greatly underestimate the emotional distress women experience from these crimes. As Professor Owen Jones has argued, the effectiveness of laws hinges on an accurate model of human nature. If that nature differs between women and men, then perhaps laws, like medicine, need to take those differences into account. Since women are the primary (although not exclusive) victims of crimes of sexual coercion, acknowledging these differences becomes doubly important. Closing the gap in legal fairness to victims may require recognizing these sex differences and perhaps creating a "reasonable woman" standard rather than a generic "reasonable person" standard for female victims.

REDUCING SEXUAL CONFLICT THROUGH THE CRUMBLING OF PATRIARCHY

It is useful to partition "patriarchy" into three distinct, albeit intertwined, levels of analysis — *institutionalized* patriarchy, patriarchal *social norms and beliefs*, and *underlying psychological adaptations* that motivate men to exert control over women. An evolutionary lens adds a fresh perspective to the analysis of these three levels subsumed by the word "patriarchy."

The clearest examples of institutionalized patriarchy center on written laws. Consider a few examples. In the United Arab Emirates, husbands and fathers, by law, have control over their wives and daughters.[9] If a woman wants to work, she must first obtain her husband's permission. She must also submit to sex with him, even against her will, except under unusual circumstances — a legal form of marital rape. A woman can be punished with one hundred lashes for having premarital sex. If a woman gets pregnant and is not married, she can be arrested; marriage is a requirement for women to have sex.

Patriarchal laws are sometimes bound up in the concept of family honor. According to one woman from the United Arab Emirates who fled Dubai for Europe and wished to remain anonymous for fear of retaliation from her home country: "Honor is a big thing in the Arab world, and family honor is within the girl — her virginity is the family's honor. If that honor is gone, the reputation of the family is gone. So, the girl has to pay the price."[10] That price is that the father and then husband have more or less exclusive control over a woman's freedom to exercise choice in when, where, and with whom she has sex. That price is danger to the woman if people perceive that her honor is lost, sometimes even death at the hands of her family members in the form of "honor killings."

These types of institutionalized patriarchy, of course, are not limited to Arab countries. In the United States, for example, "marital rape" was considered an oxymoron; the concept was incoherent. Whereas rape of non-spouses was definitely illegal, all states had *marital exception clauses*. Husbands could force sex on their wives with legal impunity. A woman who went to the authorities with a charge of marital rape would be laughed out of the police station. She had no legal recourse. It was not until 1993 that all fifty states within the United States finally deleted the marital rape exemption. Although husbands can now be charged with raping their wives, the effects of this form of institutionalized patriarchy continue to linger. As of 2019, for example, in seventeen states the husband must use physical force to qualify the act as rape.[11] Unlike for cases of non-spousal rape, he cannot be convicted of rape if the wife is unconscious, drugged, or incapacitated by means other than physical force. Attempts to remove these marital exemption clauses and impose the same legal standards for rape in marital and non-marital contexts have failed so far in the states of Ohio and Maryland. Even in states that have removed these marital loopholes, the penalties for spousal rape are often lighter than for non-marital rape. In California, for example, husbands who rape their wives, unlike other rapists, are not required to register as sex offenders unless they use physical force to commit the rape. The overall trend when it comes to laws that favor men's control over women's bodies, however, is promising. These forms of institutionalized patriarchy are fading, and continued efforts to eliminate them offer further hope of reducing sexual conflict.

Patriarchal social norms and beliefs, the second level of analysis, also show evidence of change. A study conducted in 2018 of four thousand individuals in the United Kingdom found that roughly a quarter of Brits do not consider nonconsensual marital sex to be rape, even though it legally qualifies as rape.[12] Evidence for a change in attitudes comes from dramatic differences between generations. More than 33 percent of those over the age of sixty-five, but only 16 percent of those between sixteen and twenty-four, think that forced sex within marriage does not qualify as rape. Similar cohort shifts have been found in US samples.[13] Social norms that give husbands sexual control over the bodies of their wives are changing for the better.

The #MeToo movement offers another sign of changing social norms. Men in positions of power from Hollywood to the halls of academia are being taken to task for sexual abuses. Bill Cosby, a serial sexual abuser, received a long jail sentence. The rapper R. Kelly, accused of multiple counts of sexual exploitation and abuse of girls below the age of consent, has been repeatedly denied bail while he awaits trial.[14] The proverbial "casting couch" wherein powerful Hollywood moguls extort sexual favors from aspiring actresses may soon be a distant memory. Strict policies that put those guilty of sexual crimes on a publicly accessible sexual offender registry can exploit our evolved emotion of shame, which should have positive deterrent effects. The anticipated humiliation from public knowledge will not deter all men who consider committing sexual offenses, but surely it will dissuade some.

In my own university there is required training for all employees about the rules against sexual harassment. These rules include not just quid pro quo harassment, whereby a favorable grade for a student or a promotion for an employee hinges on sexual acquiescence. They also include "hostile environment" sexual harassment, whereby sexual comments, lewd jokes, or old-fashioned boorish behavior qualify as sexual harassment. Importantly, according to a new Texas law implemented in 2020, these rules include mandatory reporting to the proper university authorities by anyone who has *witnessed* sexual harassment of one person by another. This new rule makes salient the fact that casually overlooking harassment, or dismissing it as "boys will be boys," will no longer be tolerated. The penalties are severe — one can get fired for failing to report witnessed sexual harassment. These changes have a salutary effect in shifting the burden of reporting a sexual harasser off of the sole shoulders of the victim. They render everyone a potential witness. Victims may no longer feel alone with their victimization.

Changing social norms are undoubtedly decreasing sexual abuses in the workplace. Patriarchal beliefs, that men who have attained positions of power are free to use their positions to harass or abuse those lower in the hierarchy for sexual purposes, are fading fast, along with the former lax university policies that permitted sexually inappropriate behavior. To get a sense of how dramatic changing social norms are, consider a personal example. When I was a college student, some professors threw beer keg parties in their homes for the students in their classes. At one, I witnessed a male graduate student lounging with his arms around two female undergraduate students. No one batted an eye. At each of the universities in which I have taught over the years — Berkeley, Harvard, Michigan, and Texas — it was widely known that some male professors slept with female graduate students as well as female undergraduates. Although violations of this sort undoubtedly still occur, I have not heard about a single instance within the past few years. Admittedly, these violations may still occur and are more skillfully concealed. Shifting social norms, however, seem to have followed from changes in sexual harassment policies and greater attention to the problem.

It would be a mistake to view human responsiveness to social norms as somehow separate from our evolved psychology. We are a rule-following species. A core part of our evolved psychology is to decipher social consensus, conform to group opinion, and adhere to social imperatives. Throughout human evolutionary history, people lived and died by their social reputations. Violating social rules, and especially sexual rules, brought shame to violators and sometimes reputational damage to their entire families. We care deeply about how we are perceived by others. As the evolutionary economist Robert Frank notes, "We come into this world with a nervous system that worries about rank."[15]

Plummeting in status in the eyes of others meant loss of access to the group's vital resources, including food, land, and mates.[16] People can and do pretend to follow sexual mores in public while flaunting them in private, of course. Some succeed in evading prying eyes into their private behavior. But people are also supremely adept at detecting sexual hypocrisy. Rule flaunters are often found out. And even the anxiety engendered by the prospect of being discovered changes the cost-benefit calculus in the minds of men who contemplate engaging in sexual violations to begin with.

The overall cultural trends are clear. Institutions that support male control over female bodies are eroding. The informal social customs that permitted men in positions of power to abuse with impunity are fading. Men who sexually harass women who are subordinate in the work hierarchy are getting fired. Men who rape are increasingly convicted and jailed. Just as we can create friction-free physical environments to prevent activating our skin-protecting callus-producing adaptations, we can create social environments that prevent the activation of men's more violent sexual impulses.

Patriarchy is collapsing — its institutions, its social norms, and the expression of unseemly mating mechanisms that gave rise to male advantage. But what are those mating mechanisms that led to patriarchal institutions and social norms to begin with?

And how easily can the expression of those mechanisms be suppressed?

We turn now to a third level of analysis beyond institutionalized laws and prevailing social norms that support patriarchy and explore several psychological pillars — features of men's and women's sexual psychology, and women's and men's coevolved responses to those features — that are partly responsible for the creation of patriarchal laws and social norms to begin with. These psychological pillars start with the fundamentals of mating — female mate preferences and men's strategies of mate competition.

WOMEN'S MATE PREFERENCES AND MEN'S COEVOLVED COMPETITIVE STRATEGIES

Men are one long breeding experiment run by women, according to some evolutionary scientists.[17] Men have been chosen as mates by women generation after generation to acquire and control resources. Consequently, men have evolved powerful motivations to acquire status and resources precisely because of female preferences. Men who failed to do so struggled to succeed in the competitive game of mating. Modern women have inherited these mate preferences from their successful ancestral mothers. Modern men have inherited these motivational priorities and competitive strategies from their successful ancestral fathers.

Feminist scholars rightly stress the importance of power in sexual conflict. It must be recognized, however, that men's motivations for power, status, and resources exist in part because women have preferred to mate with men who possess power, status, and resources. Neglecting this part of the causal origins of sexual conflict will impede efforts to alter it.

The importance of women's mate preferences and men's coevolved resource acquisition strategies as key causes of patriarchy may seem ironic, paradoxical, or supremely unfair. Not only do men monopolize power and resources more than women, but now women seem to be getting blamed for men's patriarchal motivations! From an evolutionary perspective, however, issues of blame are entirely beside the causal point. The key question is: If women's mate preferences have partly selected for a male psychology that places a high motivational priority on ascending the status hierarchy and acquiring resources, can these causal links in the creation of patriarchy be suppressed or overridden in ways that reduce conflict between the sexes?

One answer lies with the malleability of women's mate preferences. Women prioritize many qualities in mating besides status and resources — kindness, intelligence, reliability, emotional stability, good health, sense of humor, and adaptability. Can women elevate these preferences to supersede those of a man's resource-holding potential? Can women override their sexual attraction to powerful men, thereby severing the causal link to men's power-seeking drives? These are open questions, but I know of one woman who consciously changed her mate preferences after suffering from two abusive relationships with high-status men who felt entitled to cheat on her. She is now happily coupled with a lower-status man who treats her like a queen. It is a source of power to recognize that women hold the reins in this evolutionary equation, and their mate selections, in principle, have the power to undermine male control and create greater equality between the sexes.

A second step is the recognition that the evolutionary rationale for the origins of women's mate preferences may be largely irrelevant in modern societies. Ancestral women benefited enormously from choosing men who could provide resources that helped them and their children survive through droughts and harsh winters when food scarcity brought them to the brink of starvation. In most of the modern world, starvation has been largely eliminated. Women's mate preferences for resource-provisioning men, supremely adaptive in our evolutionary past, may have lost their warrant.

Third, most modern women simply do not need a man's resources. They can obtain their own. The fact that women in the Western world are currently outperforming men in educational achievement, from high school grades through obtaining higher professional degrees, suggests that men may soon be lagging behind women in control over resources.

It may take some time for changes in women's mate preferences to cause men to scale back on their hypercompetitive striving for status and its accoutrements. But if it is true that men are one long breeding experiment run by women over evolutionary time, and that men can adjust their forms of mate competition to embody what women want in the current environment, then there are grounds for optimism that changes in men's behavior will follow from changes in women's desires.

CURBING MALE SEXUAL POSSESSIVENESS

From an evolutionary perspective, observable cues to women's fertility are valued by men. Without sexual access to fertile women, a man and the genes within him become evolutionary dead ends. These fitness payoffs are not consciously or even unconsciously calculated; there is no evolutionary spreadsheet in men's brains that computes fitness marginal values. What evolution has produced, rather, is a male psychology that values women and is especially attracted to women showing fertility cues, and it is that psychology that gets activated and implemented in the here and now. That psychology includes adaptations that motivate men to increase their chances of success through attractive displays of status, shows of athleticism and physical prowess, the lavishing of gifts, and sometimes through deception, exploitation, or coercion.

Over evolutionary time, men who succeeded in having sex with fertile women through whatever means became ancestors. Men who failed left no descendants.

Because long-term mating, a rare mating strategy among primates, has always been critical to men's reproductive success, in men's minds mates gained often must be retained. Successful mate attraction and mate selection are not enough. Men have evolved adaptations to hold on to mates over the long term — tactics of mate retention that range from vigilance to violence. On the positive side, men retain mates by providing them with a bounty of benefits. They shower them with attention, affection, devotion, gifts, and love. They provide women with the qualities that formed the basis of their initial mate choice. On the negative side, some men retain mates using more malicious methods. They manipulate women's emotions. They monitor women's movements. They jealously guard them and curtail interactions with rival men. They sometimes strive to cut off their partner's friendships and kin ties. They threaten violence and sometimes carry out those threats. In the unfortunate and amoral evolutionary currency of differential reproductive success, men who controlled women's sexuality over the long run out-reproduced men who did not. The consequence is a male mindset of what Professors Margo Wilson and Martin Daly call *male sexual proprietariness*. As disturbing as it is, some men mistakenly think of women as they think of property — to be owned, controlled, used, and exploited.

Testaments to this male mindset go back as far as there exist written records. Consider one of the Ten Commandments, from Exodus 20:17: "You shall not covet your neighbor's house. You shall not covet your neighbor's wife, or his male or female servant, his ox or donkey, or anything that belongs to your neighbor." These are commandments to men, and wives get lumped in with other property.

For at least a couple of thousand years, a man who had sex with another man's wife was viewed as committing a property violation. The rape of a married woman by another man was viewed primarily as an offense against the husband; if unmarried, an offense against her father; if a slave, an offense against the slave owner.[18] During the 246 years of legalized slavery in the sordid history of the United States, the rape of slaves was not a criminal offense.[19] The explicit granting of legal rights to sex with the bodies of women owned by men anchors one end of this grotesquely immoral male mindset.

This male mentality is one key cause of patriarchy. It's a frame of mind that led lawmakers over centuries to write statutes granting husbands property-like rights over married women's bodies. It's a mindset that allowed men to punish women who violated those rights by refusing a husband's sexual advances. It's a mindset that leads men to acquire resources and to dole those resources out to women as a means of controlling them. And it's a mindset that causes men to use violence to prevent women from leaving and to stalk them after they have departed. These causal elements of patriarchy have a clear and unmistakable foundation in men's evolved psychology of gaining and controlling the sexuality of women.

One way to curtail men's proprietary mindset is to empower women — a trend that started with first-wave feminists who ushered in women's right to vote and continues today with women's increasing access to their own resources. Women who have their own resources are less likely to tolerate a bad marriage. They have the ability to leave when they have the resources they need for themselves and their children. This shifts the balance of power and tamps down men's proprietary mindset. Men curb their jealous control when they know women have the ability to leave. They adopt a benefit-bestowing strategy of mate retention rather than a cost-inflicting one.

A potential solution is to create a society with two key features — high levels of gender resource equality and low levels of income inequality, leveling the differences between the haves and the have-nots. Norway provides a test case.[20] It is one of the most gender-egalitarian countries in the world and provides a large safety net for everyone within its borders. Women in bad relationships are free to leave with little or no economic deprivation. Norway has one of the lowest rates of physical and sexual intimate partner violence toward women — 6 percent per year and a 27 percent lifetime prevalence.[21] These rates are still appallingly high, of course. But contrast them with another country in the developed world. Turkey is a nation far lower in gender equality and far higher in income inequality. The twelve-month rates of intimate partner physical and sexual violence toward women in Turkey are 11 percent, nearly double those in Norway, and the lifetime prevalence comes in at 38 percent.[22] The links between income inequality and intimate partner violence are also found in different regions within the same country. In India, for example, regions that have higher income inequality have higher rates of sexual violence within marriages.[23] Men at the bottom of the economic totem pole lack the resources to hold on to women who want to leave, and so some resort to coercive means to keep them.

These within- and between-cultural differences offer great promise for curtailing the violent side of men's sexual possessiveness. There is no reason to believe that the underlying psychology of male sexual jealousy differs among Norway, Turkey, India, and the United States. On the contrary, there is good evidence that Norwegians share this fundamental possessive psychology with men around the globe.[24] We cannot, of course, prove definitely that greater income equality and greater gender equality in a culture are the decisive causes of Norway's relatively low rates of men's proprietary violence toward women. Correlation does not prove causation. Nonetheless, the findings give hope that policies that decrease income inequality and increase gender equality at the cultural level have the potential to curtail men's violence toward women. And a key to enacting these policies is better representation of women in government — a factor with a solid statistical link to lower levels of partner violence.[25]

Understanding sex differences in mating psychology provides a path toward reducing violence toward women. When women have a say in designing policies around sexual violence, they are more likely to bring a female mindset to those policies — a sensibility that understands deeply the traumas that victims of sexual violence experience. And when women have the freedom and resources to leave bad relationships and men have the resources to use benefit-bestowing methods of mate retention, men are more likely to refrain from resorting to violence to hold on to a mate. Knowledge of our evolved sexual psychology can be leveraged to reduce sexual violence.

THE SEXUAL OBJECTIFICATION OF WOMEN

That men sexualize women, notice their physical appearance before many other attributes, and sometimes view women as sex objects are observations so obvious that pointing them out evokes reactions similar to noting that water is wet. Professors Tomi-Ann Roberts and Barbara Fredrickson have been at the forefront of scientists documenting evidence for the harmful effects of the sexual objectification of women.[26] They argue that media, advertisements, socialization practices, music videos, and beauty contests cause people to treat women as sex objects to the neglect of more meaningful internal qualities such as agency, character, needs, wants, feelings, and morality. Roberts and Fredrickson argue that sexual objectification leads to a host of problems for women. One is the internalization of the external — when women come to sexually objectify themselves.

Self-objectification, they argue, leads women to spend money on beauty-enhancement products, pursue potentially harmful cosmetic surgery, and feel bad about their bodies. The chronic attention to appearance enhancement detracts from the time and energy women could spend engaging in alternative activities. Internal and external sexual objectification produce a host of problems. Some center on mental health — anxiety, depression, and eating disorders. All these show profound gender differences. Women are twice as likely as men to suffer from depression, and roughly nine times as likely to experience eating disorders.[27] Some problems are work related, such as sexual harassment. And some are repeated hassles in public spaces, such as receiving catcalls, honking, ogling, leering, and comments about their breasts or buttocks.[28]

Men who view women solely as sex objects may be more likely to engage in sexual aggression. Rape victims who are dressed in sexualized attire are more likely to be unfairly blamed by others and held partly responsible for the act of a man sexually assaulting them.[29] Even absent assault, sexual objectification by men may lead men to see women as desirable for a casual sexual encounter at the expense of an enduring and fulfilling long-term mateship.

Although Roberts and Fredrickson are undoubtedly on target in their descriptions of sexual objectification and its consequences, an evolutionary perspective provides a deeper explanation for its origins, for why it occurs to begin with. From an evolutionary perspective, all living men come from a long and unbroken line of ancestral men who attended to, prioritized, and were attracted to observable physical and behavioral cues to a woman's fertility. Men who failed to do so failed to become ancestors. Physical appearance provides a wealth of probabilistically linked fertility cues — clear skin, rosy complexion, clear eyes, full lips, lustrous hair, long hair, good muscle tone, feminine facial features, feminine-sounding voices, symmetrical bodies and faces, and a low waist-to-hip ratio. Over the long expanse of human evolutionary history, men who sexualized these fertility cues, found these cues attractive, and acted on their attractions out-reproduced men who did not. Modern men, in short, have inherited adaptations that led to their ancestors' success, and evaluating women on cues to sexual attractiveness was one of the keys, however unfortunate and damaging that may be.

The priority men place on a woman's physical appearance, in turn, established one set of ground rules for female-female mate competition. Physically attractive women historically succeeded in attracting more desirable men as mates — men with status and resources who were able and willing to invest heavily in them and their children. What seems at one level of analysis to be a means for men to oppress women, a mindset of sexual objectification, has created conflict between women in the form of ferocious competition to embody what men desire in potential mates.

In the modern environment, this has resulted in a tragic form of runaway female intrasexual competition. Women are indeed profoundly affected by media images. These images hijack women's mate-competition adaptations and give them wildly distorted images of what their competitors look like. Women's actual competitors are not supermodels like Cindy Crawford or actresses like Charlize Theron. They are not the latest scantily clad social media influencers. But images of these celebrities loom large in women's minds. The use of dermal fillers in lips to make them plumper and Botox in foreheads to smooth out wrinkles is a form of female intrasexual competition. Breast implants and tummy tucks are forms of female competition. Advertisers exploit women's competitive concerns to sell products. They play on women's social comparison around appearance to sell billions of dollars of makeup each year. As more women use more beauty products and cosmetic enhancements, the actual competition accelerates among women in the appearance arena. Men's evolved sexual psychology, in short, is a key cause of women's psychology of competition with one another. The two mindsets have coevolved and are inextricably linked.

The intensity of female-female competition is further hastened by another modern invention — easy computer access to online pornography. A study of pornography consumption in the gender-egalitarian country of Denmark by 688 young adults illustrates the profound sex differences.[30] Compared to Danish women, Danish men start consuming pornography at a younger age, spend considerably more time doing so, prefer a wider range of pornography, and are more inclined to view hard-core pornography. Denmark, of course, is not unique or atypical. Similar gender differences are witnessed in all modern cultures. An endless supply of videos of attractive, naked, sexually willing women is just a computer click away. Most pornography embodies the sexual objectification of women in the extreme. It depicts women primarily as sex objects.

The easy access to an avalanche of computer pornography has amplified several negative effects on men and on women. For men, it creates distorted and unrealistic impressions of women's sexual psychology. Rapid progression from initial meeting to sex within seconds is not something most women fantasize about or desire to do in real life. The impression of hundreds of sexually attractive and accommodating women sets up wildly unrealistic expectations in men's minds. It creates conflict when real-life women feel forced to meet those expectations or to engage in an unwanted sexual performance to match what men see online. Pornography also can become the lazy man's mating. It can supplant efforts to attract women in real life, undermining goals of more deeply satisfying relationships.

The modern proliferation of porn consumption, in short, has escalated both conflict between the sexes and the harms of female-female competition. The technology behind porn access is modern. The sexual psychology it exploits is evolutionarily ancient.

The sexual objectification of women is one causal contributor, along with male sexual proprietariness, responsible for the origins and maintenance of patriarchal beliefs, values, attitudes, and behaviors suggesting that women are prized more for their sexuality than for the content of their character. This mindset, in turn, provides one causal arrow leading to negative consequences, including hiring and promotion based on physical appearance, age discrimination against women in many occupations, sexual harassment in the workplace, leering and catcalling on the streets, and unwanted touching in crowded subways.

Whether the evolved mindset that leads to these costly consequences can be short-circuited entirely is an open question, but like most psychological mechanisms, its expression in actual behavior definitely can be. Although there are currently federal laws prohibiting workplace discrimination based on extreme obesity, which is considered a disability and hence a protected category, there are no federal laws prohibiting discrimination based on physical appearance. Some jurisdictions, however, have begun this battle. The city of Santa Cruz in California has banned discrimination based on physical characteristics.[31] The District of Columbia includes physical appearance, along with sex, age, ethnicity, and sexual orientation, among the "protected categories."

It would be naïve to think that simple knowledge of the unfortunate features of male evolved mating psychology will magically curb all of its pernicious expressions. More widespread awareness of it, however, could have the beneficial effects of providing good role models and changing social norms, regardless of whether legal statutes are implemented. The Swedish students who intervened to stop Stanford's Brock Turner from sexually assaulting an unconscious woman subsequently became lauded heroes and role models who may make witnesses to sexual improprieties more likely to intervene in the future. Humans have deeply evolved concerns for status and reputation, the esteem in which they are held by others. Shunning those who discriminate based on looks, or who harass women on the street with catcalls and unwanted touching, while applauding those who put a stop to these actions, may cause some men to hold their own unsavory actions in check. It may help to curtail men's sexual objectification of women. Leveraging our evolved preoccupation with status and reputation, in short, may provide one path toward closing the psychological gaps between men and women and reducing sexual conflict.

SOLVING THE SEX RATIO PROBLEM

Sex ratio imbalances exacerbate sexual conflicts. When there exists a surplus of women relative to men on the mating market, women often feel forced to compete for desirable men by shifting to a short-term sexual strategy. This involves sexualizing their appearance and agreeing to sexual encounters after less time has elapsed before securing deeper knowledge of the content of a man's character. At the same time, men may get pushier or more aggressive about sex because they perceive that women are more likely to give in. Women in these imbalanced mating markets sometimes feel compelled to drink alcohol, smoke marijuana, or use other drugs to lower their inhibitions for engaging in sex that they might feel is unwanted during more sober moments.

The surplus of women is becoming especially acute on college campuses. At the University of Texas at Austin, where I teach, the sex ratio is 54 percent women to 46 percent men. This imbalance may not seem large at first blush. But when you do the math it translates into a hefty 17 percent more women than men in the local mating pool. So a surplus of women among educated groups caters precisely to men's short-term sexual desires because the rarer sex is always better positioned to get what they want on the mating market. The rise of hookup cultures on college campuses and of online dating sites like Tinder, AdultFriendFinder, OkCupid, and Wild is no coincidence. The initial sex ratio imbalance among educated groups gets worse for high-achieving women. They end up being forced to compete for the limited pool of educated men not just with their more numerous educated rivals, but also with less educated women whom men find desirable on other dimensions such as sexualized appearance.

One solution is to implement policies that reduce skewed sex ratios to begin with. Strange as it may seem to some, men need to catch up to women who are jetting ahead of them in many educational markets. Another is informing men and women about the effects of a surplus of women. Men inhabiting female-surplus situations may feel more sexually entitled given the range of mating choices available to them. Women sometimes become overly intoxicated, which activates the more vulnerable or exploitable components of their sexual psychology and compromises their defenses against sexually aggressive men. Educating men about these effects may reduce the numbers of female sexual victims and also prevent some men from landing on a sexual offender registry for life. At the same time, women should be informed about the predictable effects that sex ratio imbalances have on men's sexual psychology and on their own. This knowledge should help both sexes to avoid events that can create collateral damage for all involved.

TOWARD HARMONY BETWEEN THE SEXES

Combating the psychological contributors to patriarchy, such as men's sexual objectification of women and an evolved mindset that views women as property to be possessed, provides one path toward reducing sexual conflict. Eliminating sex-biased patriarchal laws and institutions provides a second path. Efforts to alter social norms, such as those that turn a blind eye to sexual coercion, provide a third. Awareness of social contexts such as sex ratio that influence mating strategies provides a fourth. All are helped by deep knowledge of sex differences in evolved mating psychology.

They are not helped by viewing women and men as identical clones in their sexual psychology.

Despite the appallingly high rates of sexual harassment, partner violence, stalking, and sexual assault, there are good grounds for optimism. The rates of physical and sexual violence within mateships have declined dramatically over spans of decades. From 1993 to 2005, for example, the rates of violence by intimate partners in the United States fell by two-thirds, perhaps because women and their allies were more likely to report it when it does occur — a dramatic shift in social norms compared to a time when most people looked the other way.[32] Similar drops in partner violence have been documented in England and Wales for the years 1995 to 2008.[33] And although some highly patriarchal countries have not seen progress this dramatic, there are positive signs in those lands as well. In nations where wife beating was once perfectly legal and seen as a husband's right, laws against it have now been put on the books, including in 25 percent of Arab states and 35 percent of sub-Saharan African countries.[34] Rates of forcible rape have also declined. In the United States, reported rape rates were 43 per 100,000 inhabitants in 1992 but fell by roughly 25 percent to 31 per 100,000 inhabitants in 2018.[35] The ideal rate is zero, of course, and rape is still dramatically underreported. Nonetheless, these trends encourage the hope that progress can continue toward the ideal goal of zero.

A second key to reducing sexual conflict resides with the recognition of profound individual differences. Despite occasional declarations to the contrary, not all men are rapists or even potential rapists. We know with great certainty that men who score high in Dark Triad traits — narcissism, Machiavellianism, and psychopathy — are much more prone to sexual coercion, especially when combined with the persistent pursuit of a short-term mating strategy. High-level Dark Triad men are more likely to deceive women on online dating sites, sexually harass women in the workplace, assault them when in relationships, and stalk them in the aftermath of breakups. In an ideal world, the minority of men who have these traits would not exist. Given that they do exist, a good strategy would be to identify them, be alert to their nefarious tactics, avoid them when possible, and ostracize or imprison them when their conduct crosses the legal line.

Detecting Dark Triad traits through facial cues is nearly impossible, but behavioral and attitudinal cues offer more promise. One is cruelty to animals. Men high in Dark Triad traits express more negative attitudes toward animals and are more likely to abuse them.[36] Before getting too deeply involved with someone, you might want to ask them how they feel about dogs and cats, or better yet watch how they react to animals. Other behavioral markers include risk-taking activities, gambling, and steep temporal discounting — prioritizing the present over delaying gratification for the future.[37] Men who score high in Light Triad traits such as empathy, honesty, and humility are least likely to be sexually coercive.[38] Identifying these profound individual differences provides an important key to avoiding the more destructive dimensions of sexual conflict.

A third key requires using an evolutionary lens to expand our focus beyond men's minds to include female minds and the co-evolution between the two. For each male offense there exists at least one female defense and typically several. We saw this with women's defenses against rape, which include alarm calling to allies through screaming, fighting to physically fend off an attacker, fleeing, seeking refuge, and, when all else fails, tonic immobility, which sometimes opens up opportunities for escape. As men strive to control women's sexuality, women have evolved strategies to evade that control. As men resort to threats and violence, women acquire resources, cultivate self-defense skills, and enlist allies. Women have evolved to be active strategists to maintain autonomy over their own bodies; women are not passive victims of men's patriarchal mindsets. Exploiting our knowledge of women's evolved sexual psychology provides an important key to progress.

A fourth key to progress resides in recognizing evolutionary mismatches and the rapid pace of cultural evolution. Women evolved in a social context with kin in close proximity and allies they could enlist to deter potential sexual predators. That historical context did not prepare women for a world of fraternity parties with spiked punch, novel date-rape drugs such as Rohypnol, and online sexual sociopaths. Consequently, cultural defenses — including laws, workplace policies, and online background checks for deception and criminality — must be invented and deployed where ancient evolved defenses no longer do the job.

Men's sexual violence toward women remains the most widespread human rights problem in the world.[39] Deep knowledge of men's and women's sexual psychology will help create conditions to reduce sexual violence. Information about the evolutionary history of sexual conflict will help. Knowledge that women are not passive pawns in a male game will help. Progress rests with the recognition of a fundamental change in sexual morality — that women themselves, not boyfriends, husbands, or fathers, should have sole autonomy over their own bodies. Female choice about when, where, with whom, and under which conditions they consent to sex is the deepest and most fundamental component of women's sexual psychology. It is a fundamental human right. Although men have coevolved strategies to undermine it, that freedom of choice should never be compromised. A deep understanding of the coevolution of sexual conflict in humans will not magically solve all problems. But I am convinced it is the light and the way.

ACKNOWLEDGMENTS

I offer tremendous thanks to friends and colleagues who generously provided detailed comments on one or more chapters in this book and the ideas contained in it: Kathy Baker, Rebecca Burch, Courtney Crosby, Patrick Durkee, Maryanne Fisher, Owen Jones, Chelsea Kilimnik, Jaimie Krems, Cindy Meston, Catherine Salmon, Donald Symons, Randy Thornhill, Griet Vandermassen, Keelah Williams, and Paula Wright. Anna Sedlacek was especially generous in providing suggestions that influenced the tone and organizational structure of several chapters. Andy Thompson provided insightful feedback on the entire manuscript. Cristine Legare provided invaluable feedback throughout the process on substance and tone. I also thank scholars whose work has influenced my thinking about sexual conflict through conversations, email exchanges, or written works: Laith Al-Shawaf, Göran Arnqvist, Kelly Asao, Mons Bendixen, April Bleske-Rechek, Kingsley Browne, Nap Chagnon, Tracey Chapman, Jaime Cloud, Dan Conroy-Beam, Leda Cosmides, Helena Cronin, Martin Daly, Richard Dawkins, Joshua Duntley, Judith Easton, A. J. Figueredo, Diana Fleischman, Steve Gangestad, Aaron Goetz, Cari Goetz, Alan Grafen, Martie Haselton, Sarah Hill, Owen Jones, L. E. Kennair, Hanna Kokko, Martin Lalumière, David Lewis, Linda Mealey, Cindy Meston,

Monique Borgerhoff Mulder, Geoffrey Parker, Carin Perilloux, Steven Pinker, Locke Rowe, Todd Shackelford, Barbara Smuts, Donald Symons, Nancy Thornhill, Randy Thornhill, John Tooby, Robert Trivers, Margo Wilson, and Richard Wrangham.

Gratitude goes to my wonderful book agent, Katinka Matson, who believed in the importance of this book from its inception. From my publisher Little, Brown Spark, Ian Straus provided insightful suggestions to the introduction and final chapter. My editor extraordinaire, Tracy Behar, brought her formidable talents to bear in spotting lapses in logic and awkward turns of phrase, and ultimately ushering this book to completion.

NOTES

CHAPTER 1: THE BATTLE OF THE SEXES

1. D. Ging, Alphas, betas, and incels: Theorizing the masculinities of the manosphere, *Men and Masculinities* (2017): 1–20, doi.org/10.1177/1097184X17706401.
2. D. M. Buss, *The Evolution of Desire: Strategies of Human Mating*, rev. and updated ed. (New York: Basic Books, 2016).
3. M. J. Albo et al., Worthless donations: Male deception and female counter play in a nuptial gift-giving spider, *BMC Evolutionary Biology* 11, no. 1 (2011): 329, https://doi.org/10.1186/1471-2148-11-329; T. Andersen et al., Why do males of the spider *Pisaura mirabilis* wrap their nuptial gifts in silk: Female preference or male control?, *Ethology* 114, no. 8 (2008): 775–781, https://doi.org/10.1111/j.1439-0310.2008.01529.x; P. E. D. Brum, L. E. Costa-Schmidt, and A. M. de Araujo, It is a matter of taste: Chemical signals mediate nuptial gift acceptance in a neotropical spider, *Behavioral Ecology* 23, no. 2 (2011): 442–447, https://doi.org/10.1093/beheco/arr209.
4. P. Seabright, *The War of the Sexes* (Princeton, NJ: Princeton University Press, 2012).
5. A. R. Pilling and C. W. M. Hart, *The Tiwi of North Australia* (New York: Holt, 1960).
6. D. M. Buss, *The Dangerous Passion: Why Jealousy Is as Necessary as Love and Sex* (New York: Free Press, 2000).
7. D. C. Queller and J. E. Strassman, Evolutionary conflict, *Annual Review of Ecology, Evolution, and Systematics* 49 (2018): 73–93.
8. W. D. Lassek and S. J. Gaulin, Costs and benefits of fat-free muscle mass in men: Relationship to mating success, dietary requirements, and native immunity, *Evolution and Human Behavior* 30, no. 5 (2009): 322–328.

NOTES

CHAPTER 2: THE MATING MARKET

1. D. P. Schmitt, Universal sex differences in the desire for sexual variety: Tests from 52 nations, 6 continents, and 13 islands, *Journal of Personality and Social Psychology* 85, no. 1 (2003): 85–104.
2. A. C. Kinsey et al., *Sexual Behavior in the Human Male* (Philadelphia: Saunders, 1953); A. C. Kinsey et al., *Sexual Behavior in the Human Female* (Philadelphia: Saunders, 1953).
3. G. Tyson et al., A first look at user activity on Tinder, in *Proceedings of the 2016 IEEE/ACM International Conference on Advances in Social Networks Analysis and Mining*, ed. R. Kumar, J. Caverlee, and H. Tong (New York: IEEE Press, 2016), 461–466.
4. J. Kasperkevic, How many Tinder users are married? Fact-checking the app's tweet storm, *Guardian*, August 12, 2015, https://www.theguardian.com/technology/2015/aug/12/tinder-vanity-fair-twitter-storm-fact-checking.
5. R. D. Clark and E. Hatfield, Gender differences in receptivity to sexual offers, *Journal of Psychology and Human Sexuality* 2, no. 1 (1989): 39–55.
6. For a review of the many studies, see D. M. Buss and D. P. Schmitt, Mate preferences and their behavioral manifestations, *Annual Review of Psychology* 70 (2019): 77–110.
7. Chris Rock, "A man is only as faithful as his options," BrainyQuote, https://www.brainyquote.com/quotes/chris_rock_378054.
8. B. Gentile et al., Gender differences in domain-specific self-esteem: A meta-analysis, *Review of General Psychology* 13, no. 1 (2009): 34–45.
9. D. T. Kenrick et al., Evolution, traits, and the stages of human courtship: Qualifying the parental investment model, *Journal of Personality* 58, no. 1 (1990): 97–116.
10. E. E. Bruch and M. E. J. Newman, Aspirational pursuit of mates in online dating markets, *Science Advances* 4, no. 8 (2018): eaap9815.
11. An oft-repeated quip by Harvard professor Irv DeVore, according to Professor Sarah Hrdy, personal communication, November 9, 2020.
12. S. E. Hill and D. M. Buss, The mere presence of opposite-sex others on judgments of sexual and romantic desirability: Opposite effects for men and women, *Personality and Social Psychology Bulletin* 34, no. 5 (2008): 635–647.
13. C. L. Toma, J. T. Hancock, and N. B. Ellison, Separating fact from fiction: An examination of deceptive self-presentation in online dating profiles,

Personality and Social Psychology Bulletin 34, no. 8 (2008): 1023–1036; R. E. Guadagno, B. M. Okdie, and S. A. Kruse, Dating deception: Gender, online dating, and exaggerated self-presentation, *Computers in Human Behavior* 28, no. 2 (2012): 642–647.

14 Cananbaum, August 31, 2013, comment on White_kong1, Have you ever had sex with someone you hate? What was it like?, Reddit, https://www.reddit.com/r/AskReddit/comments/1lgk2p/have_you_ever_had_sex_with_someone_you_hate_what/.

15 W. Von Hippel and R. Trivers, The evolution and psychology of self-deception, *Behavioral and Brain Sciences* 34, no. 1 (2011): 1–16.

16 Toma et al., Separating fact from fiction.

17 Ronin Eternales, The age and date verification scam, Online Dating Scams, December 10, 2016, https://theonlinedatingscams.com/age-or-date-verification-scam/.

18 NPR, "When Bob Packwood Was Nearly Expelled from the Senate for Sexual Misconduct," *All Things Considered,* November 27, 2017, https://www.npr.org/2017/11/27/566096392/when-bob-packwood-was-nearly-expelled-from-the-senate-for-sexual-misconduct.

19 D. M. Buss, *The Evolution of Desire: Strategies of Human Mating,* rev. and updated ed. (New York: Basic Books, 2016).

20 C. D. FitzGibbon and J. H. Fanshawe, Stotting in Thomson's gazelles: An honest signal of condition, *Behavioral Ecology and Sociobiology* 23, no. 2 (1988): 69–74.

21 B. Grayson and M. I. Stein, Attracting assault: Victims' nonverbal cues, *Journal of Communication* 31, no. 1 (1981): 68–75.

22 K. Sakaguchi and T. Hasegawa, Person perception through gait information and target choice for sexual advances: Comparison of likely targets in experiments and real life, *Journal of Nonverbal Behavior* 30, no. 2 (2006): 63–85.

23 L. S. Sugiyama, Physical attractiveness in adaptationist perspective, in *The Handbook of Evolutionary Psychology,* ed. D. M. Buss (Hoboken, NJ: Wiley, 2015), 1–68.

24 Erica Jong, in Quotable Quotes, Goodreads, https://www.goodreads.com/quotes/351729-you-see-a-lot-of-smart-guys-with-dumb-women.

25 Reis Thebault, Tucker Carlson makes sexually explicit jokes about Miss Teen USA contestant in latest audio, *Washington Post,* March 12, 2019, https://www.washingtonpost.com/arts-entertainment/2019/03/13/tucker-carlson-makes-sexually-explicit-jokes-about-miss-teen-usa-contestant-latest-audio/.

26 S. Gordon, 5 things college freshmen should know about sexual assault, Verywell Mind, October 19, 2020, https://www.verywellmind.com/what-college-freshmen-should-know-about-sexual-assault-4150032.

27 Dr. Andy Thomson, email message to the author, March 20, 2017.

28 D. Burnett, How "provocative clothes" affect the brain — and why it's no excuse for assault, *Guardian*, January 25, 2018, https://www.theguardian.com/science/brain-flapping/2018/jan/25/how-provocative-clothes-affect-the-brain-and-why-its-no-excuse-for-assault.

29 D. Ariely and G. Loewenstein, The heat of the moment: The effect of sexual arousal on sexual decision making, *Journal of Behavioral Decision Making* 19, no. 2 (2006): 87–98.

30 Ariely and Loewenstein, Heat of the moment, 94.

31 D. M. Lewis et al., Exploitative male mating strategies: Personality, mating orientation, and relationship status, *Personality and Individual Differences* 52, no. 2 (2012): 139–143.

32 M. L. Lalumière et al., *The Causes of Rape: Understanding Individual Differences in Male Propensity for Sexual Aggression* (Washington, DC: American Psychological Association, 2005).

33 C. D. Goetz, J. A. Easton, and C. M. Meston, The allure of vulnerability: Advertising cues to exploitability as a signal of sexual accessibility, *Personality and Individual Differences* 64 (2014): 121–125.

34 D. M. Buss, *The Dangerous Passion: Why Jealousy Is as Necessary as Love and Sex* (New York: Free Press, 2000).

35 Paraphrased from the song "Lyin' Eyes," music and lyrics by Don Henley and Glenn Frey (The Eagles).

36 Emicarri, Who else has dealt with unwanted attention from men due to their appearance and how have you coped with it? Needed to share my story, wanted to hear others, and get advice!, Reddit, July 27, 2018, https://www.reddit.com/r/TheGirlSurvivalGuide/comments/92fqj6/who_else_has_dealt_with_unwanted_attention_from/.

37 G. L. Carter, A. C. Campbell, and S. Muncer, The dark triad personality: Attractiveness to women, *Personality and Individual Differences* 56 (2014): 57–61.

38 Bad boys, Urban Dictionary, posted by "Peter Pan (Not using my name)," June 11, 2014, https://www.urbandictionary.com/define.php?term=Bad%20boys.

39 V. Spratt, The Dark Triad: The scientific reason why we're so attracted to fuckboys (or girls), *Grazia*, May 2, 2016, https://graziadaily.co.uk/relationships/dating/dark-triad-attracted/.

40 P. K. Jonason and D. M. Buss, Avoiding entangling commitments: Tactics

for implementing a short-term mating strategy, *Personality and Individual Differences* 52, no. 5 (2012): 606–610.

41   These items are drawn from page 31 of D. N. Jones and D. L. Paulhus, Introducing the short dark triad (SD3): A brief measure of dark personality traits, *Assessment* 21, no. 1 (2014): 28–41.

42   G. Brewer, D. De Griffa, and E. Uzun, Dark triad traits and women's use of sexual deception, *Personality and Individual Differences* 142 (2019): 42–44.

43   P. K. Jonason, N. P. Li, and D. M. Buss, The costs and benefits of the Dark Triad: Implications for mate poaching and mate retention tactics, *Personality and Individual Differences* 48, no. 4 (2010): 373–378.

CHAPTER 3: STRUGGLES WITHIN MATESHIPS

1   D. S. Judge, American legacies and the variable life histories of women and men, *Human Nature* 6 (1995): 291–323.

2   National Institute of Child Health and Human Development, How common is infertility?, last reviewed February 8, 2018, https://www.nichd.nih.gov/health/topics/infertility/conditioninfo/common.

3   P. Bogdanovich, *The Killing of the Unicorn: Dorothy Stratten (1960–1980)* (New York: William Morrow, 1984).

4   Norman Li, email message to the author, March 24, 2019.

5   J. Duntley and D. M. Buss, Backup mates (paper presented to the Annual Conference of the Human Behavior and Evolution Society, Williamsburg, VA, June 2007).

6   Names and certain details have been changed to disguise their identities.

7   Portions of this section were adapted from D. M. Buss, Why women stray: The mate-switching hypothesis, *Aeon*, October 10, 2017, https://aeon.co/essays/does-the-mate-switching-hypothesis-explain-female-infidelity.

8   S. W. Gangestad et al., Intersexual conflict across women's ovulatory cycle, *Evolution and Human Behavior* 35, no. 4 (2014): 302–308.

9   S. P. Glass and T. L. Wright, Sex differences in type of extramarital involvement and marital dissatisfaction, *Sex Roles* 12, nos. 9–10 (1985): 1101–1120.

10   H. M. Adams, V. X. Luevano, and P. K. Jonason, Risky business: Willingness to be caught in an extra-pair relationship, relationship experience, and the Dark Triad, *Personality and Individual Differences* 66 (2014): 204–207.

11   K. Hill and A. M. Hurtado, *Ache Life History: The Ecology and Demography of a Foraging People* (New York: Routledge, 2017).

12   S. O. Junare and F. M. Patel, Financial infidelity — secret saving behavior

NOTES

of the individual, *Journal of Business Management and Social Sciences Research* 1, no. 2 (2012): 40–44.

13 Junare and Patel, Financial infidelity.

14 B. T. Klontz and S. L. Britt, How clients' money scripts predict their financial behaviors, *Journal of Financial Planning* 25, no. 11 (2012): 33–43.

15 J. Dew and J. Dakin, Financial disagreements and marital conflict tactics, *Journal of Financial Therapy* 2, no. 1 (2011), https://doi.org/10.4148/jft.v2i1.1414; L. M. Papp, E. M. Cummings, and M. C. Goeke-Morey, For richer, for poorer: Money as a topic of marital conflict in the home, *Family Relations* 58, no. 1 (2009): 91–103.

16 J. Dew, S. Britt, and S. Huston, Examining the relationship between financial issues and divorce, *Family Relations* 61, no. 4 (2012): 615–628.

17 A. Coley and B. Burgess, Gender differences in cognitive and affective impulse buying, *Journal of Fashion Marketing and Management: An International Journal* 7, no. 3 (2003): 282–295.

18 A. Zahavi, Mate selection — a selection for a handicap, *Journal of Theoretical Biology* 53, no. 1 (1975): 205–214.

19 D. S. Hamermesh, *Beauty Pays: Why Attractive People Are More Successful* (Princeton, NJ: Princeton University Press, 2011).

20 J. M. Sundie et al., Peacocks, Porsches, and Thorstein Veblen: Conspicuous consumption as a sexual signaling system, *Journal of Personality and Social Psychology* 100, no. 4 (2011): 664.

21 N. S. Holtzman and M. J. Strube, People with dark personalities tend to create a physically attractive veneer, *Social Psychological and Personality Science* 4, no. 4 (2013): 461–467.

22 V. Elwin, *The Kingdom of the Young* (Oxford: Oxford University Press, 1968).

23 L. Margolin and L. White, The continuing role of physical attractiveness in marriage, *Journal of Marriage and the Family* (1987): 21–27.

24 L. Hudders et al., The rival wears Prada: Luxury consumption as a female competition strategy, *Evolutionary Psychology* 12, no. 3 (2014), https://doi.org/10.1177/147470491401200306.

25 http://www.script-o-rama.com/movie_scripts/i/i-love-you-to-death-script.html.

26 F. M. Brodie, *No Man Knows My History: The Life of Joseph Smith*, 2nd ed. (1945; New York: Knopf, 1971).

27 J. Krakauer, *Under the Banner of Heaven: A Story of Violent Faith* (New York: Anchor, 2004), 126.

28 P. Seabright, *The War of the Sexes: How Conflict and Cooperation Have*

Shaped Men and Women from Prehistory to the Present (Princeton, NJ: Princeton University Press, 2012), 26.

29 G. Gute, E. M. Eshbaugh, and J. Wiersma, Sex for you, but not for me: Discontinuity in undergraduate emerging adults' definitions of "having sex," *Journal of Sex Research* 45, no. 4 (2008): 329–337.

30 B. X. Kuhle, Did you have sex with him? Do you love her? An in vivo test of sex differences in jealous interrogations, *Personality and Individual Differences* 51, no. 8 (2011): 1044–1047.

31 J. C. Thompson, Women and the law in ancient Israel, Women in the Ancient World, n.d., http://www.womenintheancientworld.com/women%20and%20the%20law%20in%20ancient%20israel.htm.

32 M. Wilson and M. Daly, Man who mistook his wife for a chattel, in *The Adapted Mind: Evolutionary Psychology and the Generation of Culture*, ed. J. H. Barkow, L. Cosmides, and J. Tooby (New York: Oxford University Press, 1992), 289–322.

33 Wilson and Daly, Man who mistook his wife for a chattel.

34 A. Muiruri, How to use sex to punish, reward your man into behaving himself, Eve, n.d., https://www.standardmedia.co.ke/evewoman/article/2001248944/how-to-use-sex-to-punish-and-reward-your-man-into-behaving-himself.

35 C. M. Meston and D. M. Buss, *Why Women Have Sex: Understanding Sexual Motivations from Adventure to Revenge (and Everything in Between)* (New York: Macmillan, 2009), 186–187.

36 Muiruri, How to use sex.

37 B. Mustanski, How often do men and women think about sex?, *Psychology Today*, December 6, 2011, https://www.psychologytoday.com/us/blog/the-sexual-continuum/201112/how-often-do-men-and-women-think-about-sex.

38 C. P. Davis, High and low testosterone levels in men, MedicineNet, n.d., https://www.medicinenet.com/high_and_low_testosterone_levels_in_men/views.htm.

39 C. Watts et al., Withholding of sex and forced sex: Dimensions of violence against Zimbabwean women, *Reproductive Health Matters* 6, no. 12 (1998): 57–65.

40 Watts et al., Withholding of sex and forced sex, 62.

CHAPTER 4: COPING WITH RELATIONSHIP CONFLICT

1 D. M. Buss, *Evolutionary Psychology: The New Science of the Mind*, 6th ed. (New York: Routledge, 2019).

NOTES

2   B. A. Scelza et al., Patterns of paternal investment predict cross-cultural variation in jealous response, *Nature Human Behaviour* 4, no. 1 (2020): 20–26.

3   B. A. Scelza, Jealousy in a small-scale, natural fertility population: The roles of paternity, investment and love in jealous response, *Evolution and Human Behavior* 35, no. 2 (2014): 103–108.

4   Scelza, Jealousy in a small-scale, natural fertility population.

5   Scelza, Jealousy in a small-scale, natural fertility population.

6   B. X. Kuhle, Did you have sex with him? Do you love her? An in vivo test of sex differences in jealous interrogations, *Personality and Individual Differences* 51, no. 8 (2011): 1044–1047.

7   D. M. Buss et al., Distress about mating rivals, *Personal Relationships* 7, no. 3 (2000): 235–243.

8   D. M. Buss, *The Murderer Next Door: Why the Mind Is Designed to Kill* (New York: Penguin, 2006).

9   D. M. Buss, From vigilance to violence: Tactics of mate retention in American undergraduates, *Ethology and Sociobiology* 9, no. 5 (1988): 291–317; D. M. Buss and T. K. Shackelford, From vigilance to violence: Mate retention tactics in married couples, *Journal of Personality and Social Psychology* 72, no. 2 (1997): 346–361.

10 M. Wilson and M. Daly, The man who mistook his wife for a chattel, in *The Adapted Mind: Evolutionary Psychology and the Generation of Culture*, ed. J. H. Barkow, L. Cosmides, and J. Tooby (New York: Oxford University Press, 1992), 289–322.

11 Dr. Hook, "When You're in Love with a Beautiful Woman," https://www.azlyrics.com/lyrics/drhook/whenyoureinlovewithabeautifulwoman.html.

12 A. L. Bleske and T. K. Shackelford, Poaching, promiscuity, and deceit: Combatting mating rivalry in same-sex friendships, *Personal Relationships* 8, no. 4 (2001): 407–424.

13 P. K. Jonason, N. P. Li, and D. M. Buss, The costs and benefits of the Dark Triad: Implications for mate poaching and mate retention tactics, *Personality and Individual Differences* 48, no. 4 (2010): 373–378; C. J. Holden et al., Personality features and mate retention strategies: Honesty-humility and the willingness to manipulate, deceive, and exploit romantic partners, *Personality and Individual Differences* 57 (2014): 31–36. Note that low honesty-humility, the measure used in this study, is an excellent marker of high Dark Triad traits. See also W. F. McKibbin et al., Men's mate retention varies with men's personality and their partner's personality, *Personality and Individual Differences* 56 (2014): 62–67.

14. A. J. Cousins, M. A. Fugère, and M. L. Riggs, Resistance to mate guarding scale in women: Psychometric properties, *Evolutionary Psychology* 13, no. 1 (2015), https://doi.org/10.1177/147470491501300107.
15. Buss and Shackelford, From vigilance to violence.
16. M. A. Fugère, A. J. Cousins, and S. A. MacLaren, (Mis)matching in physical attractiveness and women's resistance to mate guarding, *Personality and Individual Differences* 87 (2015): 190–195.
17. J. Tooby et al., Internal regulatory variables and the design of human motivation: A computational and evolutionary approach, *Handbook of Approach and Avoidance Motivation* 15 (2008): 251.
18. A. Sell, J. Tooby, and L. Cosmides, Formidability and the logic of human anger, *Proceedings of the National Academy of Sciences* 106, no. 35 (2009): 15073–15078.
19. D. M. Buss, *The Dangerous Passion: Why Jealousy Is as Necessary as Love and Sex* (New York: Free Press, 2000).
20. G. L. White, Inducing jealousy: A power perspective, *Personality and Social Psychology Bulletin* 6, no. 2 (1980): 222–227.
21. Buss, *Dangerous Passion*; M. Daly and M. Wilson, *Homicide* (New York: Transaction Publishers, 1988); Buss, *Murderer Next Door*.
22. A. W. Delton and T. E. Robertson, How the mind makes welfare tradeoffs: Evolution, computation, and emotion, *Current Opinion in Psychology* 7 (2016): 12–16.
23. T. Leopold, Gender differences in the consequences of divorce: A study of multiple outcomes, *Demography* 55, no. 3 (2018): 769–797.
24. C. M. Meston and D. M. Buss, *Why Women Have Sex* (New York: Holt, 2009).
25. D. Conroy-Beam, C. D. Goetz, and D. M. Buss, What predicts romantic relationship satisfaction and mate retention intensity: Mate preference fulfillment or mate value discrepancies?, *Evolution and Human Behavior* 37, no. 6 (2016): 440–448.
26. Conroy-Beam et al., What predicts romantic relationship satisfaction.
27. Conroy-Beam et al., What predicts romantic relationship satisfaction.
28. J. Tooby and L. Cosmides, Friendship and the banker's paradox: Other pathways to the evolution of adaptations for altruism, in *Proceedings of the British Academy*, vol. 88, ed. W. G. Runciman, J. M. Smith, and R. I. M. Dunbar (Oxford: Oxford University Press, 1996), 119–144.

NOTES

CHAPTER 5: INTIMATE PARTNER VIOLENCE

1. https://www.un.org/press/en/2008/sgsm11437.doc.htm.
2. E. Mieure, Jury finds Matt Seals guilty on domestic violence charges, *Jackson Hole News and Guide*, June 7, 2019, https://www.jhnewsandguide.com/jackson_hole_daily/local/jury-finds-matt-seals-guilty-on-domestic-violence-charges/article_24f32353-4312-594d-b3d7-3db2046f13c6.html.
3. S. Hamby, Guess how many domestic violence offenders go to jail, *Psychology Today*, October 1, 2014, https://www.psychologytoday.com/us/blog/the-web-violence/201410/guess-how-many-domestic-violence-offenders-go-jail.
4. K. Thorstad, H. Bellino, and C. Y. Herrera, My patients are affected by intimate partner violence — and yours are, too!, Texas Academy of Family Physicians, n.d., https://www.tafp.org/news/tfp/q4-2018/IPV_hidden-epidemic.
5. U. Bacchi, Russian women post bruised selfies to push for domestic violence law, Reuters, July 24, 2019, https://www.reuters.com/article/russia-women-crime/russian-women-post-bruised-selfies-to-push-for-domestic-violence-law-idUSL8N24P38Y.
6. D. M. Buss and J. D. Duntley, The evolution of intimate partner violence, *Aggression and Violent Behavior* 16, no. 5 (2011): 411–419.
7. US District Attorney's Office, Western District of Tennessee, Federal domestic violence laws, updated May 26, 2020, https://www.justice.gov/usao-wdtn/victim-witness-program/federal-domestic-violence-laws.
8. D. G. Dutton and K. R. White, Attachment insecurity and intimate partner violence, *Aggression and Violent Behavior* 17, no. 5 (2012): 475–481.
9. K. M. Bell and A. E. Naugle, Intimate partner violence theoretical considerations: Moving towards a contextual framework, *Clinical Psychology Review* 28, no.7 (2008): 1096–1107.
10. J. Hill, *See What You Made Me Do* (Carlton, Australia: Black, 2019), 136.
11. E. S. Buzawa et al., Role of victim preference in determining police response to victims of domestic violence, in *Domestic Violence: The Changing Criminal Justice Response*, ed. E. S. Buzawa and C. G. Buzawa (Westport, CT: Auburn House, 1992).
12. A. J. Figueredo et al., Blood, solidarity, status, and honor: The sexual balance of power and spousal abuse in Sonora, Mexico, *Evolution and Human Behavior* 22, no. 5 (2001): 295–328.
13. M. Wilson and M. Daly, Lethal and nonlethal violence against wives and

the evolutionary psychology of male sexual proprietariness, *Sage Series on Violence Against Women* 9 (1998): 199–230.

14 P. Grainger, I was ashamed and I worried I wouldn't be believed about domestic violence — so I stayed longer than I should have, Journal.ie, November 21, 2018, https://www.thejournal.ie/readme/domestic-violence-ireland-4352949-Nov2018/.

15 D. Sznycer et al., Cross-cultural invariances in the architecture of shame, *Proceedings of the National Academy of Sciences* 115, no. 39 (2018): 9702–9707.

16 Wilson and Daly, Lethal and nonlethal violence.

17 As reported to the author in 2019.

18 M. Nichols, Gaslighting: What it is & how to tell if you are being gaslit, December 8, 2017, https://www.meriahnichols.com/gaslighting/.

19 J. R. Miller, Teen allegedly killed girlfriend's dog in jealous rage, dumped it on her doorstep, *New York Post*, March 15, 2019, https://nypost.com/2019/03/15/teen-allegedly-killed-girlfriends-dog-in-jealous-rage-dumped-it-on-her-doorstep/.

20 D. M. Buss, *The Murderer Next Door: Why the Mind Is Designed to Kill* (New York: Penguin, 2006).

21 Buss, *Murderer Next Door*.

22 Buss, *Murderer Next Door*.

23 D. P. Schmitt and D. M. Buss, Human mate poaching: Tactics and temptations for infiltrating existing mateships, *Journal of Personality and Social Psychology* 80, no. 6 (2001): 894–917; D. P. Schmitt, Patterns and universals of mate poaching across 53 nations: The effects of sex, culture, and personality on romantically attracting another person's partner, *Journal of Personality and Social Psychology* 86, no. 4 (2004): 560–584.

24 Wilson and Daly, Lethal and nonlethal violence.

25 M. I. Wilson and M. Daly, Male sexual proprietariness and violence against wives, *Current Directions in Psychological Science* 5, no. 1 (1996): 2–7.

26 M. Daly and M. Wilson, *Homicide* (New York: Aldine, 1988), 128.

27 D. M. Buss, *The Dangerous Passion: Why Jealousy Is as Necessary as Love and Sex* (New York: Free Press, 2000).

28 N. M. Shields and C. R. Hanneke, Attribution processes in violent relationships: Perceptions of violent husbands and their wives, *Journal of Applied Social Psychology* 13 (1983): 515–527.

29 R. L. Burch and G. G. Gallup, Pregnancy as a stimulus for domestic violence, *Journal of Family Violence* 19, no. 4 (2004): 243–247.

30 S. L. Martin et al., Changes in intimate partner violence during pregnancy, *Journal of Family Violence* 19, no. 4 (2004): 201–210; T. L. Taillieu

and D. A. Brownridge, Violence against pregnant women: Prevalence, patterns, risk factors, theories, and directions for future research, *Aggression and Violent Behavior* 15, no. 1 (2010): 14–35.

31 E. Valladares et al., Violence against pregnant women: Prevalence and characteristics. A population-based study in Nicaragua, *BJOG: An International Journal of Obstetrics and Gynecology* 112, no. 9 (2005): 1243–1248.

32 Buss, *Murderer Next Door*.

33 M. Daly, L. S. Singh, and M. Wilson, Children fathered by previous partners: A risk factor for violence against women, *Canadian Journal of Public Health* 84 (1993): 209–210.

34 M. Daly and M. Wilson, *Homicide: Foundations of Human Behavior* (New York: Routledge, 2017); M. Daly and M. Wilson, Evolutionary psychology and marital conflict: The relevance of stepchildren, in *Sex, Power, Conflict: Evolutionary and Feminist Perspectives*, ed. D. M. Buss and N. Malamuth (New York: Oxford University Press, 1996), 9–28.

35 A. Perez et al., Timing and sequence of resuming ovulation and menstruation after childbirth, *Population Studies* 25, no. 3 (1971): 491–503; Daly and Wilson, *Homicide*.

36 K. G. Anderson, H. Kaplan, and J. Lancaster, Paternal care by genetic fathers and stepfathers I: Reports from Albuquerque men, *Evolution and Human Behavior* 20, no. 6 (1999): 405–431.

37 G. C. Alexandre et al., Cues of paternal uncertainty and father to child physical abuse as reported by mothers in Rio de Janeiro, Brazil, *Child Abuse and Neglect* 35, no. 8 (2011): 567–573.

38 Buss, *Dangerous Passion*.

39 S. Rohwer, J. C. Herron, and M. Daly, Stepparental behavior as mating effort in birds and other animals, *Evolution and Human Behavior* 20, no. 6 (1999): 367–390.

40 Daly and Wilson, *Homicide*.

41 M. G. Haselton et al., Sex, lies, and strategic interference: The psychology of deception between the sexes, *Personality and Social Psychology Bulletin* 31, no. 1 (2005): 3–23.

42 D. M. Buss, The evolutionary genetics of personality: Does mutation load signal relationship load?, *Behavioral and Brain Sciences* 29, no. 4 (2006): 409.

43 D. M. Buss and T. K. Shackelford, Susceptibility to infidelity in the first year of marriage, *Journal of Research in Personality* 31, no. 2 (1997): 193–221.

44 N. Graham-Kevan and J. Archer, Control tactics and partner violence in

heterosexual relationships, *Evolution and Human Behavior* 30, no. 6 (2009): 445–452.

45 M. Wilson and M. Daly, An evolutionary psychological perspective on male sexual proprietariness and violence against wives, *Violence and Victims* 8, no. 3 (1993): 271.

46 M. C. McHugh and I. H. Frieze, Intimate partner violence: New directions, *Annals of the New York Academy of Sciences* 1087 (2006): 121; D. E. Russell, *Rape in Marriage*, exp. and rev. ed. (Bloomington: Indiana University Press, 1990).

47 C. Stanik, R. Kurzban, and P. Ellsworth, Rejection hurts: The effect of being dumped on subsequent mating efforts, *Evolutionary Psychology* 8, no. 4 (2010), https://doi.org/10.1177/147470491000800410.

48 C. Perilloux and D. M. Buss, Breaking up romantic relationships: Costs experienced and coping strategies deployed, *Evolutionary Psychology* 6, no. 1 (2008), https://doi.org/10.1177/147470490800600119.

49 J. J. Gayford, Wife battering: A preliminary survey of 100 cases, *British Medical Journal* 1 (1975): 194–197.

50 Wilson and Daly, An evolutionary psychological perspective, 281.

51 T. K. Shackelford et al., When we hurt the ones we love: Predicting violence against women from men's mate retention tactics, *Personal Relationships* 12 (2005): 447–463.

52 S. Pinker, *The Better Angels of Our Nature: Why Violence Has Declined* (New York: Penguin, 2012).

53 S. Kiire, A "fast" life history strategy affects intimate partner violence through the Dark Triad and mate retention behavior, *Personality and Individual Differences* 140 (2019): 46–51.

54 S. Kiire, Psychopathy rather than Machiavellianism or narcissism facilitates intimate partner violence via fast life strategy, *Personality and Individual Differences* 104 (2017): 401–406.

55 A. M. Mauricio, J. Tein, and F. G. Lopez, Borderline and antisocial personality scores as mediators between attachment and intimate partner violence, *Violence and Victims* 22, no. 2 (2007): 139–157.

56 E. A. Dowgwillo et al., DSM-5 pathological personality traits and intimate partner violence among male and female college students, *Violence and Victims* 31, no. 3 (2016): 416–437.

57 J. R. Peters, K. J. Derefinko, and D. R. Lynam, Negative urgency accounts for the association between borderline personality features and intimate partner violence in young men, *Journal of Personality Disorders* 31, no. 1 (2017): 16–25.

58 Borderline personality disorder, HelpGuide, https://www.helpguide.org/articles/mental-disorders/borderline-personality-disorder.htm.
59 J. J. Kreisman and H. Straus, *I Hate You — Don't Leave Me: Understanding the Borderline Personality* (New York: Penguin, 2010).
60 Wilson and Daly, Lethal and nonlethal violence.
61 Buss, *Dangerous Passion*; Wilson and Daly, Man who mistook his wife for a chattel.
62 N. A. Chagnon, *Yanomamö: The Last Days of Eden* (San Diego, CA: Harcourt Brace Jovanovich, 1992), 149.
63 A. J. Figueredo et al., Blood, solidarity, status, and honor.
64 M. Carney, F. Buttell, and D. Dutton, Women who perpetrate intimate partner violence: A review of the literature with recommendations for treatment, *Aggression and Violent Behavior* 12, no. 1 (2007): 108–115.
65 Carney et al., Women who perpetrate intimate partner violence.
66 M. A. Straus, Wife-beating: How common, and why?, *Victimology* 2 (1977–1978): 443–458.
67 R. L. McNeely and C. R. Mann, Domestic violence is a human issue, *Journal of Interpersonal Violence* 5, no. 1 (1990): 129–132.
68 R. P. Dobash et al., The myth of sexual symmetry in marital violence, *Social Problems* 39, no. 1 (1992): 71–91.
69 Dobash et al., Myth of sexual symmetry.

CHAPTER 6: STALKING AND REVENGE AFTER A BREAKUP

1 Violence against women: State and federal stalking laws, Berkman Center for Internet and Society, Online Lecture and Discussion Series, https://cyber.harvard.edu/vaw00/cyberstalking_laws.html.
2 Violence against women: State and federal stalking laws.
3 J. D. Duntley and D. M. Buss, The evolution of stalking, *Sex Roles* 66, nos. 5–6 (2012): 311–327.
4 Duntley and Buss, Evolution of stalking.
5 B. H. Spitzberg and W. R. Cupach, The state of the art of stalking: Taking stock of the emerging literature, *Aggression and Violent Behavior* 12 (2007): 64–86, https://doi.org/10.1016/j.avb.2006.05.001.
6 F. A. Hughes, K. Thom, and R. Dixon, Nature and prevalence of stalking among New Zealand mental health clinicians, *Journal of Psychosocial Nursing and Mental Health Services* 45, no. 4 (2007): 32–39.
7 D. A. Maran, A. Varetto, and M. Zedda, Italian nurses' experience of stalk-

ing: A questionnaire survey, *Violence and Victims* 29, no. 1 (2014): 109–121.

8. K. A. Roberts, Women's experience of violence during stalking by former romantic partners: Factors predictive of stalking violence, *Violence Against Women* 11, no. 1 (2005): 89–114.

9. S. D. Thomas et al., Harm associated with stalking victimization, *Australian and New Zealand Journal of Psychiatry* 42, no. 9 (2008): 800–806; Roberts, Women's experience of violence.

10. Roberts, Women's experience of violence, 101.

11. J. McFarlane et al., Stalking and intimate partner femicide, *Homicide Studies* 3 (1999): 300–316.

12. T. McEwan, P. Mullen, and R. Purcell, Identifying risk factors in stalking: A review of current research, *International Journal of Law and Psychiatry* 30 (2007): 7.

13. D. M. Buss, The evolutionary genetics of personality: Does mutation load signal relationship load?, *Behavioral and Brain Sciences* 29, no. 4 (2006): 409.

14. C. L. Patton, M. R. Nobles, and K. A. Fox, Look who's stalking: Obsessive pursuit and attachment theory, *Journal of Criminal Justice* 38, no. 3 (2010): 282–290.

15. R. C. Fraley, N. G. Waller, and K. A. Brennan, An item response theory analysis of self-report measures of adult attachment, *Journal of Personality and Social Psychology* 78, no. 2 (2000): 350–365.

16. M. Stokes, N. Newton, and A. Kaur, Stalking, and social and romantic functioning among adolescents and adults with autism spectrum disorder, *Journal of Autism and Developmental Disorders* 37, no. 10 (2007): 1969–1986.

17. J. R. Meloy, The psychology of stalking, in *The Psychology of Stalking*, ed. J. Meloy (New York: Academic Press, 1998), 1–23.

18. P. E. Mullen, M. Pathé, and R. Purcell, *Stalkers and Their Victims* (Cambridge: Cambridge University Press, 2000).

19. T. Logan et al., *Partner Stalking: How Women Respond, Cope, and Survive* (New York: Springer, 2006), 184–185.

20. For a review of these studies, see Duntley and Buss, Evolution of stalking.

21. K. K. Kienlen, Developmental and social antecedents of stalking, in Meloy, *Psychology of Stalking*, 59.

22. Kienlen, Developmental and social antecedents of stalking.

23. Kienlen, Developmental and social antecedents of stalking.

24. Kienlen, Developmental and social antecedents of stalking.

25 Mullen et al., *Stalkers and Their Victims*.
26 P. K. Durkee, A. W. Lukaszewski, and D. M. Buss, Pride and shame: Key components of a culturally universal status management system, *Evolution and Human Behavior* 40 (2019): 470–478.
27 Duntley and Buss, Evolution of stalking; J. D. Duntley and D. M. Buss, Stalking as a strategy of human mating (paper presented to the Annual Meeting of the Human Behavior and Evolution Society, New Brunswick, NJ, 2002).
28 B. S. Fisher et al., Statewide estimates of stalking among high school students in Kentucky: Demographic profile and sex differences, *Violence Against Women* 20, no. 10 (2014): 1258–1279.
29 Logan et al., *Partner Stalking*, 25.
30 T. K. Logan and R. Walker, The impact of stalking-related fear and gender on personal safety outcomes, *Journal of Interpersonal Violence* (2019): 2, http://doi.org/10.1177/0886260519829280.
31 Duntley and Buss, Evolution of stalking.
32 D. K. Citron and M. A. Franks, Criminalizing revenge porn, *Wake Forest Law Review* 49 (2014): 345.
33 S. Bates, Revenge porn and mental health: A qualitative analysis of the mental health effects of revenge porn on female survivors, *Feminist Criminology* 12, no. 1 (2017): 22–42.
34 R. Wells, The trauma of revenge porn, *New York Times*, August 4, 2019, https://www.nytimes.com/2019/08/04/opinion/revenge-porn-privacy.html.
35 Cyber Civil Rights Initiative, 46 states + DC + one territory now have revenge porn laws, n.d., https://www.cybercivilrights.org/revenge-porn-laws/; Wikipedia, s.v. Revenge porn, last updated October 23, 2020, 21:18, https://en.wikipedia.org/wiki/Revenge_porn.
36 A. Powell, A. Flynn, and N. Henry, 1 in 5 Australians is a victim of "revenge porn," despite new laws to prevent it, Conversation, July 22, 2019, https://theconversation.com/1-in-5-australians-is-a-victim-of-revenge-porn-despite-new-laws-to-prevent-it-117838.
37 https://www.msn.com/en-au/news/world/it-was-as-if-he-pressed-the-trigger-of-a-gun-father-of-revenge-porn-victim-21-who-killed-herself-after-her-boyfriend-blackmailed-her-says-she-was-cruelly-betrayed/ar-AAFjMHm.
38 M. Pathé, *Surviving Stalking* (Cambridge: Cambridge University Press, 2002); Mullen et al., *Stalkers and Their Victims*.

NOTES

CHAPTER 7: SEXUAL COERCION

1. I thank these authors for the cancer analogy: W. F. McKibbin, T. K. Shackelford, A. T. Goetz, and V. G. Starratt, Why do men rape? An evolutionary psychological perspective, *Review of General Psychology* 12, no. 1 (2008): 86–97.
2. O. D. Jones and T. H. Goldsmith, Law and behavioral biology, *Columbia Law Review* 105 (2005): 405–502; O. D. Jones, Sex, culture, and the biology of rape: Toward explanation and prevention, *California Law Review* 87 (1999): 827.
3. J. K. Maner et al., Can't take my eyes off you: Attentional adhesion to mates and rivals, *Journal of Personality and Social Psychology* 93, no. 3 (2007): 389.
4. L. Penke, Revised sociosexual orientation inventory, in *Handbook of Sexuality-Related Measures*, ed. T. D. Fisher et al. (New York: Routledge, 2011), 622–625.
5. I. Aharon et al., Beautiful faces have variable reward value: fMRI and behavioral evidence, *Neuron* 32, no. 3 (2001): 537–551.
6. J. J. Palamar and M. Griffin, Non-consensual sexual contact at electronic dance music parties, *Archives of Sexual Behavior* (2020): 1–9.
7. S. Prescott-Smith, Two in five young female festival goers have been subjected to unwanted sexual behaviour, YouGov, June 21, 2018, https://yougov.co.uk/topics/lifestyle/articles-reports/2018/06/21/two-five-young-female-festival-goers-have-been-sub.
8. C. Mellgren, M. Andersson, and A.-K. Ivert, "It happens all the time": Women's experiences and normalization of sexual harassment in public space, *Women and Criminal Justice* 28 (2018): 262–291, https://doi.org/10.1080/08974454.2017.1372328.
9. Sexual harassment is a "major public health issue" in Sweden: Survey, the Local, May 28, 2019, https://www.thelocal.se/20190528/sexual-harassment-is-a-major-public-health-issue-in-sweden-survey.
10. A. X. Estrada and A. W. Berggren, Sexual harassment and its impact for women officers and cadets in the Swedish Armed Forces, *Military Psychology* 21, no. 2 (2009): 162–185; E. Witkowska and E. Menckel, Perceptions of sexual harassment in Swedish high schools: Experiences and school-environment problems, *European Journal of Public Health* 15, no. 1 (2005): 78–85.
11. K. Hill and A. M. Hurtado, *Ache Life History: The Ecology and Demography of a Foraging People* (New York: Routledge, 2017).

NOTES

12 L. H. Keeley, *War Before Civilization* (New York: Oxford University Press, 1997).

13 S. B. Hrdy, *Mothers and Others* (Cambridge, MA: Harvard University Press, 2011); R. L. Burch, More than just a pretty face: The overlooked contributions of women in evolutionary psychology textbooks, *Evolutionary Behavioral Sciences* 14, no. 1 (2020): 100.

14 Chart: The percentage of women and men in each profession, *Boston Globe*, March 6, 2017, https://www.bostonglobe.com/metro/2017/03/06/chart-the-percentage-women-and-men-each-profession/GBX22YsWl0XaeHghwXfE4H/story.html.

15 R. A. Clay, Women outnumber men in psychology, but not in the field's top echelons, American Psychological Association, July–August 2017, https://www.apa.org/monitor/2017/07-08/women-psychology.

16 D. Conroy-Beam and D. M. Buss, Why is age so important in human mating? Evolved age preferences and their influences on multiple mating behaviors, *Evolutionary Behavioral Sciences* 13, no. 2 (2019): 127–157.

17 S. Kanazawa and M. C. Still, Teaching may be hazardous to your marriage, *Evolution and Human Behavior* 21, no. 3 (2000): 185–190.

18 U.S. Equal Employment Opportunity Commission, Sexual Harassment, n.d., https://www.eeoc.gov/laws/types/sexual_harassment.cfm.

19 M. J. Gelfand, L. F. Fitzgerald, and F. Drasgow, The structure of sexual harassment: A confirmatory analysis across cultures and settings, *Journal of Vocational Behavior* 47, no. 2 (1995): 164–177.

20 What we learned from the Harvey Weinstein documentary Untouchable, BBC News, August 31, 2019, https://www.bbc.com/news/entertainment-arts-49522102.

21 Gelfand et al., Structure of sexual harassment.

22 R. Jones, Women report sexual harassment more than men, even in male-dominated workplaces, *Ms.*, August 7, 2018, https://msmagazine.com/2018/08/07/women-report-sexual-harassment-men-even-male-dominated-workplaces/.

23 L. Stemple, A. Flores, and I. H. Meyer, Sexual victimization perpetrated by women: Federal data reveal surprising prevalence, *Aggression and Violent Behavior* 34 (2017): 302–311.

24 M. V. Studd and U. E. Gattiker, The evolutionary psychology of sexual harassment in organizations, *Ethology and Sociobiology* 12, no. 4 (1991): 249–290.

25 J. A. Bargh et al., Attractiveness of the underling: An automatic power → sex association and its consequences for sexual harassment and aggression, *Journal of Personality and Social Psychology* 68, no. 5 (1995): 768–781.

26 L. F. Fitzgerald, *The Last Great Open Secret: The Sexual Harassment of Women in the Workplace and Academia* (Washington, DC: Federation of Behavioral, Psychological and Cognitive Sciences, 1993).

27 K. R. Browne, Sex, power, and dominance: The evolutionary psychology of sexual harassment, *Managerial and Decision Economics* 27, nos. 2–3 (2006): 145–158.

28 Nicholas Fraser, *Aristotle Onassis* (New York: Lippincott, 1977), 308.

29 D. M. Buss, Conflict between the sexes: Strategic interference and the evocation of anger and upset, *Journal of Personality and Social Psychology* 56, no. 5 (1989): 735–747.

30 Studd and Gattiker, Evolutionary psychology of sexual harassment.

31 D. E. Terpstra and S. E. Cook, Complainant characteristics and reported behaviors and consequences associated with formal sexual harassment charges, *Personnel Psychology* 38, no. 3 (1985): 559–574.

32 Terpstra and Cook, Complainant characteristics.

33 B. A. Gutek, *Sex and the Workplace* (San Francisco: Jossey-Bass, 1985), 46.

34 Browne, Sex, power, and dominance.

35 L. Klümper and S. Schwarz, Oppression or opportunity? Sexual strategies and the perception of sexual advances, *Evolutionary Psychological Science* (2019): 1–12.

36 Studd and Gattiker, Evolutionary psychology of sexual harassment.

37 J. Semmelroth and D. Buss, unpublished data.

38 S. L. Ream, When service with a smile invites more than satisfied customers: Third-party sexual harassment and the implications of charges against Safeway, *Hastings Women's Law Journal* 11 (2000): 107–122; Browne, Sex, power, and dominance.

39 E. Steel, How Bill O'Reilly silenced his accusers, *New York Times*, April 4, 2018, https://www.nytimes.com/2018/04/04/business/media/how-bill-oreilly-silenced-his-accusers.html.

40 Wendy Walsh says Bill O'Reilly "became hostile" after she rebuffed his alleged sexual advances, ABC News, April 4, 2017, https://abcnews.go.com/US/wendy-walsh-bill-oreilly-hostile-rebuffed-alleged-sexual/story?id=46568606.

41 M. Bendixen and L. E. O. Kennair, Advances in the understanding of same-sex and opposite-sex sexual harassment, *Evolution and Human Behavior* 38, no. 5 (2017): 583–591.

42 K. Lee, M. Gizzarone, and M. C. Ashton, Personality and the likelihood to sexually harass, *Sex Roles* 49, nos. 1–2 (2003): 59–69.

NOTES

43 V. Zeigler-Hill et al., The Dark Triad and sexual harassment proclivity, *Personality and Individual Differences* 89 (2016): 47–54.

44 M. Noureddine et al., Analysis of courtship success in the funnel-web spider *Agelenopsis aperta*, *Behaviour* 137, no. 1 (2000): 93–117.

45 Quote about Harvey Weinstein from alleged sexual abuse victim, from the Hulu-produced documentary *Untouchable*.

46 M. P. Ghiglieri, *The Dark Side of Man: Tracing the Origins of Male Violence* (New York: Perseus, 1999).

47 L. DeMause, War as righteous rape and purification, *Journal of Psychohistory* 27, no. 4 (2000): 356; L. Zimmer-Tamakoshi, Rape and other sexual aggression, in *Encyclopedia of Sex and Gender: Men and Women in the World's Cultures* (New Haven, CT: Yale University Press, 2003), 230–243; E. Biocca, *Yanoáma: The Narrative of a White Girl Kidnapped by Amazonian Indians* (New York: Dutton, 1970).

48 H. Flynn, K. Cousins, and E. N. Picciani, Tinder lets known sex offenders use the app. It's not the only one, BuzzFeed News, December 2, 2019, https://www.buzzfeednews.com/article/hillaryflynn/tinder-lets-known-sex-offenders-use-the-app-its-not-the.

49 Parts of this section have been adapted from D. M. Buss, *The Evolution of Desire: Strategies of Human Mating*, rev. and updated ed. (New York: Basic Books, 2016), chap. 7.

50 Studies reviewed by M. L. Lalumière et al., *The Causes of Rape: Understanding Individual Differences in Male Propensity for Sexual Aggression* (Washington, DC: American Psychological Association, 2005).

51 N. M. Malamuth, Predictors of naturalistic sexual aggression, *Journal of Personality and Social Psychology* 50, no. 5 (1986): 953–962.

52 R. Farrow, *Catch and Kill* (New York: Little, Brown, 2019).

53 L. Mealey, Combating rape: Views of an evolutionary psychologist, in *Evolutionary Psychology and Violence: A Primer for Policymakers and Public Policy Advocates*, ed. R. W. Bloom and N. Dess (Westport, CT: Praeger, 2003), 83–113.

54 Mealey, Combating rape, 91.

55 M. W. Kraus, S. Côté, and D. Keltner, Social class, contextualism, and empathic accuracy, *Psychological Science* 21, no. 11 (2010): 1716–1723.

56 L. L. Betzig, *Despotism and Differential Reproduction: A Darwinian View of History* (New York: Aldine, 1986). For documentation of priests and religious leaders using power for forced sex, see D. M. Buss, Sex, marriage, and religion: What adaptive problems do religious phenomena solve?, *Psychological Inquiry* 13, no. 3 (2002): 201–203.

57 J. Krakauer, *Under the Banner of Heaven: A Story of Violent Faith* (New

NOTES

York: Anchor, 2004); N. Carlisle, Polygamist leader Warren Jeffs is accused of horrific sex abuse — again — and people on the Utah-Arizona line may have to pay — again, *Salt Lake Tribune*, January 11, 2018, https://www.sltrib.com/news/polygamy/2018/01/10/polygamist-leader-warren-jeffs-is-accused-of-horrific-sex-abuse-again-and-people-on-the-utah-arizona-line-may-have-to-pay-again/.

58 B. Jacobs, S. Siddiqui, and S. Bixby, "You can do anything": Trump brags on tape about using fame to get women, *Guardian*, October 8, 2016, https://www.theguardian.com/us-news/2016/oct/07/donald-trump-leaked-recording-women.

59 Lalumière et al., *Causes of Rape*.

60 C. Y. Senn et al., Predicting coercive sexual behavior across the lifespan in a random sample of Canadian men, *Journal of Social and Personal Relationships* 17, no. 1 (2000): 95–113.

61 R. Jewkes et al., Rape perpetration by young, rural South African men: Prevalence, patterns and risk factors, *Social Science and Medicine* 63, no. 11 (2006): 2949–2961.

62 S. Williams, *Invisible Darkness: The Strange Case of Paul Bernardo and Karla Homolka* (New York: Bantam, 2009).

63 R. Thornhill and N. W. Thornhill, Human rape: An evolutionary analysis, *Ethology and Sociobiology* 4, no. 3 (1983): 137–173.

64 R. B. Felson and P. R. Cundiff, Age and sexual assault during robberies, *Evolution and Human Behavior* 33, no. 1 (2012): 10–16.

65 D. Symons, *The Evolution of Human Sexuality* (New York: Oxford University Press, 1979), 284.

66 S. Brownmiller, *Against Our Will: Men, Women, and Rape* (New York: Ballantine, 1975).

67 N. M. Malamuth, Rape proclivity among males, *Journal of Social Issues* 37, no. 4 (1981): 138–157.

68 R. K. Young and D. Thiessen, The Texas rape scale, *Ethology and Sociobiology* 13, no. 1 (1992): 19–33.

69 C. Bijleveld, A. Morssinkhof, and A. Smeulers, Counting the countless: Rape victimization during the Rwandan genocide, *International Criminal Justice Review* 19, no. 2 (2009): 208–224.

70 Bijleveld et al., Counting the countless, 216.

71 Bijleveld et al., Counting the countless, 220.

72 J. J. Lehmiller, *Tell Me What You Want: The Science of Sexual Desire and How It Can Help You Improve Your Sex Life* (New York: Da Capo Lifelong Books, 2018).

73 A handful of studies support this contention. One example is A. D. Smith, Aggressive sexual fantasy in men with schizophrenia who commit contact sex offences against women, *Journal of Forensic Psychiatry* 10, no. 3 (1999): 538–552.

74 C. Crépault and M. Couture, Men's erotic fantasies, *Archives of Sexual Behavior* 9, no. 6 (1980): 565–581.

75 H. Leitenberg and K. Henning, Sexual fantasy, *Psychological Bulletin* 117, no. 3 (1995): 469.

76 V. Greendlinger and D. Byrne, Coercive sexual fantasies of college men as predictors of self-reported likelihood to rape and overt sexual aggression, *Journal of Sex Research* 23, no. 1 (1987): 1–11.

77 Greendlinger and Byrne, Coercive sexual fantasies.

78 Greendlinger and Byrne, Coercive sexual fantasies.

79 D. Lisak and P. M. Miller, Repeat rape and multiple offending among undetected rapists, *Violence and Victims* 17, no. 1 (2002): 73–84.

80 S. Newman, Why men rape, *Aeon*, March 30, 2017, https://aeon.co/essays/until-we-treat-rapists-as-ordinary-criminals-we-wont-stop-them.

81 R. Jewkes et al., Prevalence of and factors associated with non-partner rape perpetration: Findings from the UN Multi-country Cross-Sectional Study on Men and Violence in Asia and the Pacific, *Lancet Global Health* 1, no. 4 (2013): e208–e218.

82 Lalumière et al., *Causes of Rape*.

83 J. Freemont, Rapists speak for themselves, in *The Politics of Rape: The Victim's Perspective*, ed. D. E. H. Russell (New York: Stein and Day, 1975), 244–246.

84 E. O'Connor, In their words: The Swedish heroes who caught the Stanford sexual assailant, BuzzFeed News, June 8, 2016, https://www.buzzfeednews.com/article/emaoconnor/meet-the-two-swedish-men-who-caught-brock-turner.

85 M. R. Burt, Cultural myths and supports for rape, *Journal of Personality and Social Psychology* 38, no. 2 (1980): 217–230.

86 R. F. Baumeister, K. R. Catanese, and H. M. Wallace, Conquest by force: A narcissistic reactance theory of rape and sexual coercion, *Review of General Psychology* 6, no. 1 (2002): 92–135.

87 N. M. Malamuth, Criminal and noncriminal sexual aggressors: Integrating psychopathy in a hierarchical-mediational confluence model, *Annals of the New York Academy of Sciences* 989, no. 1 (2003): 33–58.

88 L. Penke and J. B. Asendorpf, Beyond global sociosexual orientations: A more differentiated look at sociosexuality and its effects on courtship and

romantic relationships, *Journal of Personality and Social Psychology* 95, no. 5 (2008): 1113–1135.

89 G. G. Abel et al., Self-reported sex crimes of nonincarcerated paraphiliacs, *Journal of Interpersonal Violence* 2, no. 1 (1987): 3–25.

90 M. R. Weinrott and M. Saylor, Self-report of crimes committed by sex offenders, *Journal of Interpersonal Violence* 6, no. 3 (1991): 286–300.

91 Lisak and Miller, Repeat rape.

92 R. L. Burch and G. G. Gallup Jr., Abusive men are driven by paternal uncertainty, *Evolutionary Behavioral Sciences* (2019).

93 M. Hale, *Historia Placitorum Coronae: The History of the Pleas of the Crown* (London: Gyles, Woodward, and Davis, 1736).

94 Wikipedia, s.v. Marital rape laws by country, last updated September 23, 2020, 15:48, https://en.wikipedia.org/wiki/Marital_rape_laws_by_country.

95 E. K. Martin, C. T. Taft, and P. A. Resick, A review of marital rape, *Aggression and Violent Behavior* 12, no. 3 (2007): 329–347.

96 Martin et al., Review of marital rape.

97 K. C. Basile, Prevalence of wife rape and other intimate partner sexual coercion in a nationally representative sample of women, *Violence and Victims* 17 (2002): 511–524.

98 Lalumière et al., *Causes of Rape*.

99 D. M. Buss et al., The mate switching hypothesis, *Personality and Individual Differences* 104 (2017): 143–149; Martin et al., Review of marital rape.

100 D. M. Buss, *The Dangerous Passion: Why Jealousy Is as Necessary as Love and Sex* (New York: Free Press, 2000).

101 Martin et al., Review of marital rape.

102 A. Van der Straten et al., Sexual coercion, physical violence, and HIV infection among women in steady relationships in Kigali, Rwanda, *AIDS and Behavior* 2, no. 1 (1998): 61–73.

103 R. Jewkes et al., Rape perpetration.

104 P. K. Jonason, M. Girgis, and J. Milne-Home, The exploitive mating strategy of the Dark Triad traits: Tests of rape-enabling attitudes, *Archives of Sexual Behavior* 46, no. 3 (2017): 697–706.

105 S. Pinker, *The Blank Slate* (New York: Viking, 2002), 163.

106 N. Diamond-Smith and K. Rudolph, The association between uneven sex ratios and violence: Evidence from 6 Asian countries, *PLoS One* 13, no. 6 (2018): e0197516.

107 S. Pinker, *The Better Angels of Our Nature: Why Violence Has Declined* (New York: Viking, 2011).

NOTES

CHAPTER 8: DEFENDING AGAINST SEXUAL COERCION

1. J. Hartung, *Biblical Roots of the Long Leash on Men*, unpublished manuscript (n.d.).
2. Genghis Khan, quoted in T. Royle, *A Dictionary of Military Quotations* (New York: Simon and Schuster Children's Publishing, 1989).
3. C. Palmer, Is rape a cultural universal? A re-examination of the ethnographic data, *Ethnology* 28, no. 1 (1989): 1–16.
4. T. Gregor, *Anxious Pleasures: The Sexual Lives of an Amazonian People* (Chicago: University of Chicago Press, 1985).
5. N. A. Chagnon, Life histories, blood revenge, and warfare in a tribal population, *Science* 239, no. 4843 (1988): 985–992.
6. S. Brownmiller, *Against Our Will: Men, Women, and Rape* (New York: Ballantine, 1975).
7. I. Chang, *The Rape of Nanking: The Forgotten Holocaust of World War II* (New York: Basic Books, 2014).
8. H. Sinnreich, "And it was something we didn't talk about": Rape of Jewish women during the Holocaust, *Holocaust Studies* 14, no. 2 (2008): 1–22.
9. B. Allen, *Rape Warfare: The Hidden Genocide in Bosnia-Herzegovina and Croatia* (Minneapolis: University of Minnesota Press, 1996).
10. T. Cooper, *Great Lakes Conflagration: Second Congo War, 1998–2003* (Solihull, UK: Helion and Company, 2013).
11. P. R. Sanday, The socio-cultural context of rape: A cross-cultural study, *Journal of Social Issues* 37, no. 4 (1981): 5–27.
12. Palmer, Is rape a cultural universal?, 3.
13. F. Salesius, *Diekirolneu-Insel Jap* (Berlin, 1906), translated for the Human Relations Area Files, 117.
14. R. Karsten, *Indian Tribes of the Argentine and Bolivian Chaco*, Societas Scientiarum Fennica Commentationes Humanarum Litterarium, vol. 4 (1932), 62. Quoted in Palmer, Is rape a cultural universal?
15. C. S. Larsen, Bioarchaeology, in *The International Encyclopedia of Biological Anthropology*, ed. Wenda Trevathan (Hoboken, NJ: Wiley, 2018), 1–14; C. P. Larsen, *Bioarchaeology: Interpreting Behaviour from the Human Skeleton*, 2nd ed. (Cambridge: Cambridge University Press, 2015).
16. M. Potts and T. Hayden, *Sex and War: How Biology Explains Warfare and Terrorism and Offers a Path to a Safer World* (Dallas, TX: BenBella Books, 2010).
17. T. Zerjal et al., The genetic legacy of the Mongols, *American Journal of Human Genetics* 72 (2003): 717–721.

NOTES

18. L. T. Moore et al., A Y-chromosome signature of hegemony in Gaelic Ireland, *American Journal of Human Genetics* 78, no. 2 (2006): 334–338; B. McEvoy et al., The scale and nature of Viking settlement in Ireland from Y-chromosome admixture analysis, *European Journal of Human Genetics* 14, no. 12 (2006): 1288.
19. S. Goodacre et al., Genetic evidence for a family-based Scandinavian settlement of Shetland and Orkney during the Viking periods, *Heredity* 95, no. 2 (2005): 129–135.
20. T. C. Zeng, A. J. Aw, and M. W. Feldman, Cultural hitchhiking and competition between patrilineal kin groups explain the post-Neolithic Y-chromosome bottleneck, *Nature Communications* 9, no. 1 (2018): 1–12.
21. D. M. Buss, Conflict between the sexes: Strategic interference and the evocation of anger and upset, *Journal of Personality and Social Psychology* 56, no. 5 (1989): 735–747.
22. Texas candidate's comment about rape causes a furor, *New York Times*, March 26, 1990, https://www.nytimes.com/1990/03/26/us/texas-candidate-s-comment-about-rape-causes-a-furor.html.
23. D. Russell, *The Politics of Rape: The Victim's Perspective* (New York: Stein and Day, 1975).
24. D. M. Buss, *The Evolution of Desire: Strategies of Human Mating*, rev. and updated ed. (New York: Basic Books, 2016).
25. M. Daly and M. Wilson, *Homicide: Foundations of Human Behavior* (New York: Routledge, 2017).
26. D. Mukamana and P. Brysiewicz, The lived experience of genocide rape survivors in Rwanda, *Journal of Nursing Scholarship* 40, no. 4 (2008): 382. The Interahamwe were and are a Hutu paramilitary organization that perpetrated many of the rape and genocidal atrocities against the Tutsi, Twa, and others in the Congo and Uganda in 1994. Wikipedia, s.v. Interahamwe, last updated September 30, 2020, 11:32, https://en.wikipedia.org/wiki/Interahamwe.
27. H. Liebling, H. Slegh, and B. Ruratotoye, Women and girls bearing children through rape in Goma, Eastern Congo: Stigma, health and justice responses, *Itupale Online Journal of African Studies* 4 (2012): 18–44.
28. Mukamana and Brysiewicz, Lived experience of genocide rape survivors, 382.
29. United Nations OCHA, Raped and rejected: Women face double tragedy in the Democratic Republic of the Congo, November 25, 2017, https://medium.com/humanitarian-dispatches/raped-and-rejected-women-face-double-tragedy-in-the-democratic-republic-of-the-congo-aa97adbf1552.

30 C. Perilloux, J. D. Duntley, and D. M. Buss, The costs of rape, *Archives of Sexual Behavior* 41, no. 5 (2012): 1104.

31 M. L. Lalumière et al., *The Causes of Rape: Understanding Individual Differences in Male Propensity for Sexual Aggression* (Washington, DC: American Psychological Association, 2005).

32 Mukamana and Brysiewicz, Lived experience of genocide rape survivors, 382.

33 Perilloux et al., Costs of rape, 1102.

34 Perilloux et al., Costs of rape, 1103.

35 M. Van Egmond et al., The relationship between sexual abuse and female suicidal behavior, *Crisis: The Journal of Crisis Intervention and Suicide Prevention* 14, no. 3 (1993): 129–139.

36 P. A. Resick, The psychological impact of rape, *Journal of Interpersonal Violence* 8, no. 2 (1993): 223–255.

37 The criminal justice system: Statistics, Rainn, n.d., https://www.rainn.org/statistics/criminal-justice-system.

38 United Nations OCHA, Raped and rejected.

39 M. Wilson and S. L. Mesnick, An empirical test of the bodyguard hypothesis, in *Feminism and Evolutionary Biology*, ed. P. A. Gowaty (Boston: Springer, 1997), 505–511.

40 For summaries of the relevant scientific studies, see Buss, *Evolution of Desire*.

41 D. E. Brown and Y. Chia-yun, "Big man" in universalistic perspective, unpublished manuscript, Department of Anthropology, University of California at Santa Barbara (1993).

42 Wilson and Mesnick, Empirical test of the bodyguard hypothesis.

43 C. Perilloux, J. D. Duntley, and D. M. Buss, Susceptibility to sexual victimization and women's mating strategies, *Personality and Individual Differences* 51, no. 6 (2011): 783–786.

44 A. L. Bleske-Rechek and D. M. Buss, Opposite-sex friendship: Sex differences and similarities in initiation, selection, and dissolution, *Personality and Social Psychology Bulletin* 27, no. 10 (2001): 1310–1323.

45 R. Weitz, Vulnerable warriors: Military women, military culture, and fear of rape, *Gender Issues* 32, no. 3 (2015): 172.

46 C. Perilloux, D. S. Fleischman, and D. M. Buss, The daughter-guarding hypothesis: Parental influence on, and emotional reactions to, offspring's mating behavior, *Evolutionary Psychology* 6, no. 2 (2008): 217–233.

47 A. J. Figueredo et al., Blood, solidarity, status, and honor: The sexual balance of power and spousal abuse in Sonora, Mexico, *Evolution and Human Behavior* 22, no. 5 (2001): 295–328.

48 W. F. McKibbin, Development and initial psychometric assessment of the rape avoidance inventory, *Personality and Individual Differences* 46, no. 3 (2009): 336–340.

49 M. Warr, Fear of rape among urban women, *Social Problems* 32, no. 3 (1985): 238–250.

50 B. Softas-Nall, A. Bardos, and M. Fakinos, Fear of rape: Its perceived seriousness and likelihood among young Greek women, *Violence Against Women* 1, no. 2 (1995): 174–186.

51 Warr, Fear of rape among urban women.

52 R. R. Dobbs, C. A. Waid, and T. O. C. Shelley, Explaining fear of crime as fear of rape among college females: An examination of multiple campuses in the United States, *International Journal of Social Inquiry* 2, no. 2 (2009).

53 I. S. Marks, The development of normal fear: A review, *Journal of Child Psychology and Psychiatry* 28, no. 5 (1987): 667–697.

54 Daly and Wilson, *Homicide*.

55 S. Griffin, Rape: The all-American crime, *Ramparts* September (1971): 26.

56 G. Fishman and G. S. Mesch, Fear of crime in Israel: A multidimensional approach, *Social Science Quarterly* (1996): 76–89; G. Gangoli, Controlling women's sexuality: Rape law in India, *International Approaches to Rape* (2011): 101–120.

57 J. J. M. Van Dijk, Public attitudes toward crime in the Netherlands, *Victimology* 3, no. 3 (1978): 265–273.

58 K. F. Ferraro, *Fear of Crime: Interpreting Victimization Risk* (Albany: State University of New York Press, 1995); Warr, Fear of rape among urban women.

59 E. Suarez and T. M. Gadalla, Stop blaming the victim: A meta-analysis on rape myths, *Journal of Interpersonal Violence* 25, no. 11 (2010): 2010–2035.

60 R. B. Felson, Blame analysis: Accounting for the behavior of protected groups, *American Sociologist* 22, no. 1 (1991): 5–23.

61 L. Heath and L. Davidson, Dealing with the threat of rape: Reactance or learned helplessness, *Journal of Applied Social Psychology* 18 (1988): 1334–1351.

62 M. T. Gordon and S. Riger, *The Female Fear* (New York: Free Press, 1989).

63 D. W. Pryor and M. R. Hughes, Fear of rape among college women: A social psychological analysis, *Violence and Victims* 28, no. 3 (2013): 443–465.

NOTES

64 C. M. Hilinski, Fear of crime among college students: A test of the shadow of sexual assault hypothesis, *American Journal of Criminal Justice* 34, nos. 1–2 (2009): 84–102.

65 A. Poropat, Understanding women's rape experience and fears (doctoral dissertation, University of Queensland, 1992).

66 Weitz, Vulnerable warriors, 169.

67 A. Holgate, Sexual harassment as a determinant of women's fear of rape, *Australian Journal of Sex, Marriage and Family* 10, no. 1 (1989): 21–28.

68 Holgate, Sexual harassment as a determinant, 23.

69 Warr, Fear of rape among urban women.

70 B. S. Fisher, F. T. Cullen, and M. Turner, *The Sexual Victimization of College Women* (Washington, DC: National Institute of Justice and Bureau of Justice Statistics, 2000).

71 S. E. Hickman and C. L. Muehlenhard, College women's fears and precautionary behaviors relating to acquaintance rape and stranger rape, *Psychology of Women Quarterly* 21, no. 4 (1997): 527–547.

72 Pryor and Hughes, Fear of rape among college women, 456.

73 C. C. Ward, *Attitudes Toward Rape: Feminist and Social Psychological Perspectives* (Thousand Oaks, CA: Sage, 1995).

74 D. M. Buss, *The Murderer Next Door: Why the Mind Is Designed to Kill* (New York: Penguin, 2006).

75 Pryor and Hughes, Fear of rape among college women.

76 G. de Becker, *The Gift of Fear* (Boston: Little, Brown, 1997).

77 C. Cantor, *Evolution and Posttraumatic Stress: Disorders of Vigilance and Defence* (New York: Routledge, 2005).

78 This table draws inspiration from Chapter 16 of M. Del Giudice, *Evolutionary Psychopathology: A Unified Approach* (New York: Oxford University Press, 2018).

79 Buss, *Murderer Next Door*.

80 J. M. Zoucha-Jensen and A. Coyne, The effects of resistance strategies on rape, *American Journal of Public Health* 83, no. 11 (1993): 1633–1634.

81 P. B. Bart and P. H. O'Brien, Stopping rape: Effective avoidance strategies, *Signs: Journal of Women in Culture and Society* 10, no. 1 (1984): 83–101; V. L. Quinsey and D. Upfold, Rape completion and victim injury as a function of female resistance strategy, *Canadian Journal of Behavioural Science/Revue Canadienne des Sciences du Comportement* 17, no. 1 (1985): 40–50; G. Kleck and S. Sayles, Rape and resistance, *Social Problems* 37, no. 2 (1990): 149–162.

82 S. E. Ullman, A 10-year update of "review and critique of empirical studies

of rape avoidance," *Criminal Justice and Behavior* 34 (2007): 411–429.
83 D. J. Stevens, Predatory rape avoidance, *International Review of Modern Sociology* (1994): 108.
84 Stevens, Predatory rape avoidance, 109.
85 Stevens, Predatory rape avoidance.
86 S. E. Ullman, Reflections on researching rape resistance, *Violence Against Women* 20, no. 3 (2014): 343–350.
87 Del Giudice, *Evolutionary Psychopathology*.
88 B. P. Marx et al., Tonic immobility as an evolved predator defense: Implications for sexual assault survivors, *Clinical Psychology: Science and Practice* 15, no. 1 (2008): 74–90.
89 T. Fusé et al., Factor structure of the Tonic Immobility Scale in female sexual assault survivors: An exploratory and confirmatory factor analysis, *Journal of Anxiety Disorders* 21, no. 3 (2007): 265–283; Marx et al., Tonic immobility as an evolved predator defense.
90 G. Galliano et al., Victim reactions during rape/sexual assault: A preliminary study of the immobility response and its correlates, *Journal of Interpersonal Violence* 8, no. 1 (1993): 110.
91 Russell, *Politics of Rape*, 233.
92 Marx et al., Tonic immobility as an evolved predator defense.
93 Galliano et al., Victim reactions during rape/sexual assault, 110.
94 K. D. Suschinsky and M. L. Lalumière, Prepared for anything? An investigation of female genital arousal in response to rape cues, *Psychological Science* 22, no. 2 (2011): 159–165; R. J. Levin and W. van Berlo, Sexual arousal and orgasm in subjects who experience forced or non-consensual sexual stimulation — a review, *Journal of Clinical Forensic Medicine* 11, no. 2 (2004): 82–88.
95 Suschinsky and Lalumière, Prepared for anything?
96 M. J. Bovin et al., Tonic immobility mediates the influence of peritraumatic fear and perceived inescapability on posttraumatic stress symptom severity among sexual assault survivors, *Journal of Traumatic Stress* 21, no. 4 (2008): 402–409; A. Möller, H. P. Söndergaard, and L. Helström, Tonic immobility during sexual assault — a common reaction predicting post-traumatic stress disorder and severe depression, *Acta Obstetricia et Gynecologica Scandinavica* 96, no. 8 (2017): 932–938.
97 W. D. Allen, The reporting and underreporting of rape, *Southern Economic Journal* (2007): 623–641; E. A. Lehner, Rape process templates: A hidden cause of the underreporting of rape, *Yale Journal of Law and Feminism* 29 (2017): 207.

98 A. Myhill and J. Allen, *Rape and Sexual Assault of Women: Findings from the British Crime Survey*, Home Office Research Study 159 (London: Home Office, 2002), 1–6.
99 C. Miller, *Know My Name: A Memoir* (New York: Penguin, 2019).
100 Many studies document the link beween sexual assault and PTSD; A. Elklit and D. M. Christiansen, ASD and PTSD in rape victims, *Journal of Interpersonal Violence* 25, no. 8 (2010): 1470–1488.
101 B. O. Rothbaum et al., A prospective examination of post-traumatic stress disorder in rape victims, *Journal of Traumatic Stress* 5, no. 3 (1992): 455–475.
102 C. Cantor, Post-traumatic stress disorder: Evolutionary perspectives, *Australian and New Zealand Journal of Psychiatry* 43, no. 11 (2009): 1038–1048.
103 Cantor, Post-traumatic stress disorder.
104 K. A. Chivers-Wilson, Sexual assault and posttraumatic stress disorder: A review of the biological, psychological and sociological factors and treatments, *McGill Journal of Medicine* 9, no. 2 (2006): 111.
105 Chivers-Wilson, Sexual assault and posttraumatic stress disorder.

CHAPTER 9: MINDING THE SEX GAP

1 S. Casaus, Is Ghislaine Maxwell now being held in solitary confinement?, Film Daily, August 17, 2020, https://filmdaily.co/news/ghislaine-maxwell-solitary-confinement/.
2 L. Al-Shawaf, D. M. Lewis, and D. M. Buss, Sex differences in disgust: Why are women more easily disgusted than men?, *Emotion Review* 10, no. 2 (2018): 149–160; C. L. Crosby et al., Six dimensions of sexual disgust, *Personality and Individual Differences* 156 (2020): 109714.
3 M. Mercao, Over half of millennial women have received unsolicited NSFW pics, Bustle, October 11, 2017, https://www.bustle.com/p/how-many-women-have-received-dick-pics-according-to-research-over-half-of-millennial-women-have-2893328.
4 D. Nations, What is AirDrop? How does it work?, Lifewire, October 23, 2020, https://www.lifewire.com/what-is-airdrop-how-does-it-work-1994512.
5 A. North, One state has banned unsolicited dick pics. Will it fix the problem?, Vox, September 3, 2019, https://www.vox.com/policy-and-politics/2019/9/3/20847447/unsolicited-dick-pics-texas-law-harassment.

NOTES

6. E. Hunt, "20 minutes of action": Father defends Stanford student son convicted of sexual assault, *Guardian*, June 5, 2016, https://www.theguardian.com/us-news/2016/jun/06/father-stanford-university-student-brock-turner-sexual-assault-statement.

7. Chief Compliance Officer and Vice President for Diversity and Community Engagement, Handbook of Operating Procedures 3-3031: Prohibition of sexual assault, interpersonal violence, stalking, sexual harassment, and sex discrimination, University of Texas at Austin, last reviewed August 12, 2020, https://policies.utexas.edu/policies/prohibition-sex-discrimination-sexual-harassment-sexual-assault-sexual-misconduct.

8. A. Jackson et al., Gender differences in response to lorazepam in a human drug discrimination study. *Journal of Psychopharmacology* 19, no. 6 (November 2005), 614–619.

9. Wikipedia, s.v. Human rights in the United Arab Emirates, last updated October 26, 2020, 20:58, https://en.wikipedia.org/wiki/Human_rights_in_the_United_Arab_Emirates.

10. V. Grigoriadis, The flight of the princess, *Vanity Fair*, March 2020, 128.

11. M. Pauly, It's 2019, and states are still making exceptions for spousal rape, *Mother Jones*, November 21, 2019, https://www.motherjones.com/crime-justice/2019/11/deval-patrick-spousal-rape-laws/.

12. When did marital rape become a crime?, *Week*, December 6, 2018, https://www.theweek.co.uk/98330/when-did-marital-rape-become-a-crime.

13. C. Ferro, J. Cermele, and A. Saltzman, Current perceptions of marital rape: Some good and not-so-good news, *Journal of Interpersonal Violence* 23, no. 6 (2008): 764–779.

14. G. Kaufman, Why has R. Kelly been in jail for the past year?, *Billboard*, August 13, 2020, https://www.billboard.com/articles/business/legal-and-management/9430230/r-kelly-why-in-jail-for-year.

15. R. Frank, *Choosing the Right Pond: Human Behavior and the Quest for Status* (New York: Oxford University Press, 1985).

16. D. M. Buss et al., Human status criteria: Sex differences and similarities across 14 countries, *Journal of Personality and Social Psychology* (2020).

17. Paraphrase of a quip often stated by the late Harvard professor Irv Devore, according to Professor Sarah Hrdy, personal communication, November 9, 2020.

18. S. Pinker, *The Better Angels of Our Nature: Why Violence Has Declined* (New York: Penguin, 2012).

19. R. A. Feinstein, *When Rape Was Legal: The Untold History of Sexual Violence During Slavery* (New York: Routledge, 2018).

NOTES

20 M. Bendixen, L. E. O. Kennair, and D. M. Buss, Jealousy: Evidence of strong sex differences using both forced choice and continuous measure paradigms, *Personality and Individual Differences* 86 (2015): 212–216.

21 UN Global Database on Violence Against Women, Norway, https://evaw global-database.unwomen.org/en/countries/europe/norway.

22 UN Global Database on Violence Against Women, Turkey, https://evaw global-database.unwomen.org/es/countries/asia/turkey.

23 A. S. Rashada and M. F. Sharaf, Income inequality and intimate partner violence against women: Evidence from India (Frankfurt School — Working Paper Series no. 222, Frankfurt School of Finance and Management, Frankfurt am Main, 2016).

24 Bendixen et al., Jealousy.

25 J. Archer, Does sexual selection explain human sex differences in aggression?, *Behavioral and Brain Sciences* 32, nos. 3–4 (2009): 249–266.

26 B. L. Fredrickson and T. A. Roberts, Objectification theory: Toward understanding women's lived experiences and mental health risks, *Psychology of Women Quarterly* 21, no. 2 (1997): 173–206; T. A. Roberts, R. M. Calogero, and S. J. Gervais, Objectification theory: Continuing contributions to feminist psychology, in APA *Handbook of the Psychology of Women: History, Theory, and Battlegrounds*, ed. C. B. Travis et al. (Washington, DC: American Psychological Association, 2018), 249–271.

27 M. M. Weissman et al., Sex differences in rates of depression: Cross-national perspectives, *Journal of Affective Disorders* 29, nos. 2–3 (1993): 77–84; M. Altemus, N. Sarvaiya, and C. N. Epperson, Sex differences in anxiety and depression clinical perspectives, *Frontiers in Neuroendocrinology* 35, no. 3 (2014): 320–330; M. A. Katzman, S. A. Wolchik, and S. L. Braver, The prevalence of frequent binge eating and bulimia in a nonclinical college sample, *International Journal of Eating Disorders* (1984).

28 H. B. Kozee et al., Development and psychometric evaluation of the interpersonal sexual objectification scale, *Psychology of Women Quarterly* 31, no. 2 (2007): 176–189.

29 S. Loughnan et al., Sexual objectification increases rape victim blame and decreases perceived suffering, *Psychology of Women Quarterly* 37, no. 4 (2013): 455–461.

30 G. M. Hald, Gender differences in pornography consumption among young heterosexual Danish adults, *Archives of Sexual Behavior* 35, no. 5 (2006): 577–585.

31 J. J. McDonald Jr., "Lookism": The next form of illegal discrimination, *Bloomberg Law Reports — Labor and Employment* 4, no. 46 (2010), https

NOTES

://www.fisherphillips.com/media/publication/5565_26956_fisher_phillips_mcdonald_article.pdf.

32 Pinker, *Better Angels of Our Nature*, 410–411.
33 Pinker, *Better Angels of Our Nature*, 412.
34 Pinker, *Better Angels of Our Nature*, 413.
35 Statista Research Department, Reported forcible rape rate in the United States from 1990 to 2019, Statista, September 28, 2020, https://www.statista.com/statistics/191226/reported-forcible-rape-rate-in-the-us-since-1990/.
36 P. S. Kavanagh, T. D. Signal, and N. Taylor, The Dark Triad and animal cruelty: Dark personalities, dark attitudes, and dark behaviors, *Personality and Individual Differences* 55, no. 6 (2013): 666–670.
37 L. C. Crysel et al., The dark triad and risk behavior, *Personality and Individual Differences* 54, no. 1 (2013): 35–40.
38 S. B. Kaufman et al., The light vs. dark triad of personality: Contrasting two very different profiles of human nature, *Frontiers in Psychology* 10 (2019): 467.
39 Pinker, *Better Angels of Our Nature*.

INDEX

abandonment, fear of, 62, 92–93, 115, 120, 137, 139–40
abortion, 131, 132, 206
Absher (app), 23–24
academic environments, 173, 258–59
Ache, 79
active signaling, 54
adorned attractiveness, 82–83
adultery laws, 88
AdultFriendFinder, 272
Afghanistan, 230
Against Our Will (Brownmiller), 211
age factors
   in intimate partner violence, 140–41
   in sexual harassment, 173–74, 177, 204
Agelenopsis aperta, 181
aggressive behavior, 120, 139, 198, 214–15
agreeableness, 52–53
AirDrop, 252
Aka, 173
Albania, 202
alcohol, 50–51, 148–49, 170, 182, 186, 231, 237, 272, 273
Alder, Shannon L., 92
alloparenting, 173
"alpha chads," 8
amygdala, 52, 227
anger, 93, 108, 148
animals, sex with, 196
anterior cingulate cortex, 52
antisocial personality disorder, 139

anxiety, 51, 120, 148, 151, 159
appeasement, as defensive strategy against sexual assault, 239
Arapesh, 211
Ariely, Dan, 52
ASD (autism spectrum disorder), 151
Ashley Madison (dating website), 33–34
attachment disorders, 120
attachment theory, 150–51
attention, unwanted sexual, 11
attention seeking, 54
attentional adhesion, 168–70, 204
attentive immobility, 234–35
attractiveness
   adorned vs. unadorned, 82–83
   and attentional adhesion, 168–70
   discrepancies of, in sexual marketplace, 36–38
   and female intrasexual competition, 269–70
   sexual vs. long-term-mate, 49
auditory looming bias, 234
Australia, 31, 121, 161–62, 184, 202, 212
Austria, 34
autism spectrum disorder (ASD), 151
auto mechanics, 173
avoidance, in PTSD, 245
avoidant attachment style (defense against mate guarding), 106
avoiding contact (tactic against mate guarding), 104

INDEX

backup mates, 66–69, 113
"bad boy" paradox, 55–58
"baggage" from previous relationships, 113–14
Ban Ki-moon, 118
Bangladesh, 198
bank accounts, secret, 80
Barbados, 202
Bargh, John, 176
battered women, 122, 125–26, 130, 133, 137
"battle of the sexes," 7
Bergman, Ingrid, 127
Bernardo, Paul, 191–92
"betas," 8, 11
biases of perception. *See* perception biases
Bible, 197
bidirectional domestic violence, 122
birds, 61, 63, 203
birth control pill, 24, 25
black widow spiders, 11
blaming the victim, 241, 243–44, 268
Bleske-Rechek, April, 224–25
bodyguard hypothesis, 222–26
bodyguards, sexual, 5, 48, 138, 163, 177, 184, 235, 238
Bogdanovich, Peter, 64
Bolivia, 96
borderline personality disorder (BPD), 139–40, 151
Bosnia and Herzegovina, 211
Botox, 269
Botswana, 70, 210
BPD (borderline personality disorder), 139–40, 151
brain, 52, 169–70, 227
Brazil, 27, 95, 141, 184, 202, 210–11
breakups, 151–53, 203, 219
breast implants, 269
breast milk, 16
breastfeeding, 70, 133, 173
bribery, sexual, 178
brick masons, 173
Brownmiller, Susan, 193, 211
Buffett, Warren, 82
burial sites, 212
The Burning Bed (film), 143

California, 145, 146, 257–58
California State University, San Bernardino, 76
Cambodia, 198
Canada, 120, 133, 175, 191
Cantor, Chris, 245, 246
Carlson, Tucker, 49
Carroll, Lewis, 22
cars, 39, 81, 82, 164, 165
Carter, Gregory, 55
Carter, Jimmy, 35
Carter, Rosalynn, 35
Cassell, Carol, 7
"casting couch," 258
casual sex, 5, 33, 37, 38, 50, 82, 105, 160, 161, 169, 177–80
catfishing, 40–41
cell phones, 23, 104
Chagnon, Napoleon, 223
chastity belts, 23, 141
cheating. *See* infidelity
cheetahs, 10, 45–46
childcare, 173
children, 62–64, 113, 216–17
Chile, 95
China, 95, 198, 206, 211, 229
choice, power of, 28
Cinderella story, 72
claustration, 141
clitoridectomy, 23
clothing, as cue, 51–52
Cobain, Kurt, 166
coevolution, sexual conflict, 10
cognitive ability, 48, 49
college campuses, 51
colostrum, 16
committed relationships. *See* mateships
concealment, in aftermath of sexual assault, 243–44
concrete workers, 173
conflict. *See* relationship conflict; sexual conflict
Conroy-Beam, Daniel, 76
conspicuous consumption, 81, 83–84
control, intimate partner violence as form of, 121
controlling men, 104–5
copyright law, 161
Cosby, Bill, 7, 181, 189–90, 258
cosmetics, 81, 269
couple infertility, 63
courtship displays, 10
Cousins, Alita, 103–5

INDEX

covert tactics (against mate guarding), 103
COVID-19 pandemic, 138
Crawford, Cindy, 269
Crenshaw, Kimberlé, 248
Croatia, 211
Crow Creek, 212
cryptic female choice, 15
cultural arms races, 25–26
cultural evolution, 22–26
cyberstalking, 146, 164

Dallas, Texas, 122
Daly, Martin, 130, 137, 143, 264
dance venues, unwanted sexual attention at, 170
Dark Triad, 5, 29
    and adorned attractiveness, 82–83
    and "bad boy" paradox, 55–58
    and evoking jealousy, 110
    and intimate partner violence, 138–41
    and mate guarding, 101–2
    and rape, 198–200
    and sexual exploitability, 52–53
    and sexual harassment, 181, 205
    and strategies for reducing sexual conflict, 274–75
    in women, 58–60, 78
dating apps. *See* online dating sites and apps
de Becker, Gavin, 208, 233
death of partner, 63
death threats, 128, 137
deception, 38–41
Declaration of Independence, 36
deflection tactics, 251
Democratic Republic of Congo, 211
Denmark, 24, 34, 121, 270
depression, 151, 159, 268
dermal fillers, 269
dick pics, 252
"direct benefits," 57
disgust, sexual, 251–53
District of Columbia, 271
ditziness, 54
division of labor, 172
divorce, 28, 63, 64, 74, 77, 80, 90, 136–37, 219

DNA, 15, 16, 24, 213
documentation, of stalking behavior, 164
domestic violence, 119–20, 122. *See also* intimate partner violence (IPV)
dopamine, 51
double standards, sexual, 84–87
Downs, Diane, 134
Dr. Hook, 100
drug use, illegal, 149
dual-mating strategy, 71
ducks, 203
Duggar, Josh, 33
Duntley, Joshua, 66, 154, 162, 182

Eagles (rock band), 55
Easter Island, 212
Easton, Judith, 53
eating disorders, 268
ecstasy (drug), 51, 170
Ecuador, 96
educating men, about inappropriate sexual behavior, 252–53
educational achievement, 263
eggs, 4, 15
emotional abuse, 127–28
emotional infidelity, 87
emotional manipulability, 47–48
empathy, 53, 199
endorphins, 50–51
England, 88, 95
Epstein, Jeffrey, 8, 249
equal rights, 36
eunuchs, 141
European Union, 121
evasive action, 237–38
evolutionary mismatch, 183
evolutionary perspective, 4, 5
    on Dark Triad traits, 57
    on intimate partner violence, 121–23, 132
    on jealousy, 93
    limitations and value of, 248–49
    on mate switching, 70–73
    on mating harmony, 62–66
    on rape, 180, 183–84, 186–93
    on resource infidelity, 80–81
    on sexual coercion, 166–68, 172–73, 175–77, 215–16

INDEX

evolutionary perspective (*cont.*)
   on sexual treachery, 38
   on stalking, 146–47
exogamy, 27
exploitability, 45–53
extraversion, 47, 54
eye contact, 151, 179

Facebook, 82, 164
families, effect of rape on, 218
family honor, 256–57
fantasies and fantasizing, 89–90, 195–96
fatigue, as cue, 48, 50
Fawcett, Farrah, 143
FDA (Food and Drug Administration), 24
fear
   of abandonment, 62, 92–93, 115, 120, 137, 139–40
   as emotion, 227–28
   of rape, 182–85, 226–34
   stranger, 185, 228, 232
female-dominated professions, 173
feminism (feminist theory), 16, 176, 215, 228, 233, 261, 265
fertility (fertility cues), 47, 75, 187, 193, 263–64, 268–69
fetus-murder hypothesis, 131–32
fight-or-flight response, 227–28
fighting, against sexual assault, 238–39
Figueredo, A. J., 124, 141
Fiji, 96
financial conflicts, 80–81
financial resources, 126–27, 149
financial threats, 128
Finland, 24, 121
Fisher, R. A., 57
flashbacks, 244, 245
flirting (flirtatiousness), 48, 51, 54, 75
   evoking jealousy with, 110
   as harassment, 174, 179
   and mate guarding, 101–2
Florida, 34
fMRI (functional magnetic resonance imaging), 169–70
Food and Drug Administration (FDA), 24
foot binding, 141
forceful resistance, to sexual assault, 238–39
forgiveness, 110–11

Fox News, 179
Frank, Robert, 260
Franken, Al, 7
Fredrickson, Barbara, 267
"freezing," 234–35, 240
friends and friendships, 100–101, 111–12, 142
fruit flies, 11
functional magnetic resonance imaging (fMRI), 169–70
funnel-web spider, 181
future harms, threat of, 128

gametes, 4
Gaslight (film), 127
gaslighting, 127–28
gay men, 32
gazelles, 10, 45–46, 99
Gebusi, 173, 184
gender, 4
gender resource inequality, 266
gender roles, traditional, 120, 180
genital mutilation, 141
genital photos, decontextualized, 251–52
Germany, 211
ghosting, 40
gift giving, 145
The Gift of Fear (de Becker), 233
Glamour, 38–39
Goetz, Cari, 52, 53, 76
good genes hypothesis, 71
government, women in, 267
Grant, Hugh, 35
Greece, 226, 237
Gregor, Thomas, 210
Griffin, Susan, 228–29
Guatemala, 228
gullibility, 48

Hadza, 96
handicap principle, 81
harassers, serial, 28–29
harassment, 172–81
Harvard University, 205, 259
height, 222–23
heroism, acts of, 65
High Machs, 29
hijacking, of victim's psychology, 122–28
Himba, 96, 97

INDEX

Hollywood, 258
Holocaust, 211
home break-ins, fear of, 229
home healthcare providers, 173
homicide, 98, 143, 149
Homolka, Karla, 192
honesty-humility (personality trait), 180–81
Hopi, 228
"hostile environment" sexual harassment, 259
hostile masculinity, 199
houses, expensive, 39
Hughes, Francine, 143
Hughes, James, 143
Huichol, 211
humiliation, 153–54
hunter-gatherers, 18, 88, 172
Hurley, Elizabeth, 35
Hutu, 195
hypercontrolling men, 104–5

I Hate You — Don't Leave Me (Kreisman and Straus), 139
I Love You to Death (film), 84
Iceland, 121
illegal drug use, 149
immaturity, as cue, 49–50
incapacitation, cues of, 48
incels, 7, 11, 36
incest, 194
income, exaggeration of, 39
income inequality, 266
India, 83, 206, 229, 266
Indonesia, 96, 198
infant bestowal, 17–18
infanticide, 206, 228
inferential biases, 42–45
infertility, 63, 216, 242
infibulation, 23
infidelity
  among Tiwi, 19–21
  as dangerous activity, 77
  and desire for variety, 31–36
  emotional, 87
  intimate partner violence and threat of, 130, 131
  jealousy and risk of, 93–95
  and marital rape, 203
  in mateships, 70–79
  men's vs. women's definitions of, 86–87
  and rape, 219
  resource, 79–84
insects, 11
insecure attachment disorder, 120
insecure attachment styles, 150–51
Instagram, 164
institutionalized patriarchy, 256–58
intentionally evoked jealousy, 109–10
intergroup warfare, 172
internet dating. *See* online dating sites and apps
intimate partner violence (IPV), 77, 98, 118–44
  bidirectional, 122
  and Dark Triad, 138–41
  declining rates of, 274
  evolutionary perspective on, 121–23, 132
  and hijacking of victim's psychology, 122–28
  and legal structures, 118–21
  and mate poaching, 129–30
  men's motives for perpetrating, 129–38
  in patriarchal societies, 120, 121
  women's defenses against, 141–44
intoxication, as cue, 48, 50–51
Inuit, 70
investment, relationship, 105–6, 134–35
IPV. *See* intimate partner violence
Ireland, 95, 213
irreplaceability
  creating, 114–17
  of lost partners, from stalkers' view, 155
  and marital rape, 203
Islamic cultures, 88
Israel (ancient), 87–88
Israel (modern), 181, 229
Italy, 148

Jackson, Wyoming, 118–19
Japan, 32, 138–39, 211
jealousy, 22–25, 92–98, 109–10
Jews, 211
Jonason, Peter, 57–58
Jones, Owen, 167–68, 255–56
Jong, Erica, 49
Jossipalenya, Damilya, 162

INDEX

Karo Batak, 96
Kelly, R., 258
Kentucky, 154
Khan, Genghis, 210, 212–13
Kinsey, Alfred, 32, 70
kissing, 86
Kline, Kevin, 84
Know My Name (Miller), 254
Koger's Island, 212
!Kung San, 70, 210, 228

labor, division of, 172
lactation, 16
Lalumière, Martin, 191
Lauer, Matt, 7
laws and policies
   on adultery, 88
   improving, 254–56
   on intimate partner violence, 118–21
   and patriarchy, 8
   on sexting, 252
   on sexual coercion, 167–68
   on sexual harassment, 174
   on spousal rape, 201–2
   on stalking, 145–46
   and tonic immobility, 241
   on wife beating, 274
learned behaviors, 120, 121
Lebanon, 31
lesbian women, 32
Lewis, David, 52
life-long love, myth of, 72
Light Triad, 52, 275
lions, 135
Loewenstein, George, 52
lorazepam, 255
low cognitive ability, 48, 49
Lucky (Sebold), 239, 254

Mace, 164, 237
Machiavellianism, 5, 29, 55–56, 59, 83
Mackey, John, 82
Maimonides, 210
Malamuth, Neil, 188
Malaysia, 210
male-dominated professions, 173
male sexual over-perception bias, 42–44, 171, 179, 251
male sexual proprietariness, 264

mandatory reporting, 259
Maner, Jon, 168
manosphere, 7, 8
marijuana, 51
marital rape, 201–4, 256–58
marital status, and sexual victimization, 224
Maryland, 257
Mataco, 211
Match, 140
mate-choice copying effect, 39
mate-deprivation hypothesis, 189–92
mate guarding, 99–106
   and intimate partner violence, 129–30, 140–41
   resistance to, in women, 102–6
   and stalking, 152
"Mate-Guarding Model" of rape, 202–3
mate poachers and poaching, 100–101, 103, 107, 129–30
mate-retention tactics, 83–84, 123–24, 264
mate-switching hypothesis, 71–79
mate value, 64–65, 73–76, 81–82
   discrepancies in, 93–94, 98, 107–11, 135–36
   maintaining one's, 115
   and stepchildren, 134–35
   and woman's mating decisions, 215–16
mateships, 61–91
   "backup mates" in, 66–69
   evolutionary perspective on, 62–66
   intimate partner violence as means of sustaining, 136–37
   reasons for affairs in, 70–79
   resource infidelity in, 79–84
   sexual double standards in, 84–87
   sexual withdrawal/bestowal in, 87–91, 108
mating markets, 28, 30–60
   attractiveness discrepancies in, 36–38
   and "bad boy" paradox, 55–58
   Dark Triad traits in, 55–60
   deception and treachery in, 38–41
   and desire for sexual variety, 31–36
   exploitability in, 45–53
   perception biases in, 42–45
   signaling in, 45–46, 53–55
   strategic interference on, 30–31
mating strategies, differences in, 3
Maxim, 38
Maxwell, Ghislaine, 249

INDEX

Mayanga, 96
Mbuti, 211
Mead, Margaret, 211
Mealey, Linda, 189, 190, 202–3, 216
men. *See also* sex differences
   attentional adhesion in, 168–70
   conspicuous consumption by, 81
   controlling, 104–5
   and desire for sexual variety, 31–36, 204
   educating about inappropriate sexual behavior, 252–53
   height and torso shape of, as desirable characteristics, 222–23
   infidelity among, 77
   infidelity as defined by, 86
   intimate partner violence against, 122
   mate guarding in, 100–101
   and mate value, 98
   motives of, for perpetrating intimate partner violence, 129–38
   physical size and strength of, 27
   self-esteem in, 37
   sex differences between women and, 4–5, 15–16
   sexual jealousy in, 23–25, 97
   sexual possessiveness of, 263–67
   as sexual predators, 28–29
   as stalking victims, 147
   under-perception bias in, 43–44
men's bodies, and reproduction, 16
men's magazines, 39
Mesnick, Sarah, 216, 222–24
Meston, Cindy, 53, 89
#MeToo movement, 7, 258
Mexico, 124, 212
Middle East, 23–24, 31, 229
military, sexual harassment/assault in the, 171, 225, 230
Miller, Chanel, 243, 253, 254
Miss Teen USA, 49
money, 39, 80, 177
monitoring, 23
monkey branching, 106, 113
monogamy, 62
Mormon Church, 84–85, 190–91
mugging, 46
murder, 98, 149, 182–83, 194, 227
The Murderer Next Door (Buss), 128
Muria, 83
muscle mass, 27

Namibia, 96
narcissism, 5, 29, 43–44, 55, 59, 83, 108, 135, 199–200
narcissistic personality disorder, 139, 151
negative reinforcement contingency, 158
Netherlands, 34, 97, 229, 231
neuroticism, 47
New York City, 46, 80, 170, 252
New York State, 161
New Zealand, 31
Newman, Paul, 35
Niall, King, 213
Nicaragua, 96, 132
non-WEIRD cultures, 96–97
Nordic countries, 24, 212, 213
Norway, 13, 95, 180, 266
nucleus accumbens, 169–70, 204
numbers, power of, 27

Oceania, 31
offspring, 62
Ohio, 257
OkCupid, 185, 272
olfactory cues, 234
Onassis, Aristotle, 177
online dating sites and apps, 11–12, 25–26, 38, 39, 41, 53–54, 116, 140, 185, 249, 272–73
online pornography, 249, 270–71
openness to experience, 54
opportunity costs, 21–22, 158
oral-genital contact, 85–86
O'Reilly, Bill, 7, 179
over-perception biases, sexual, 42–44, 171, 179, 251
ovulation, 15, 75, 133

Packwood, Bob, 43
Palmer, Craig, 187
Papua New Guinea, 173, 184, 198
Paraguay, 79
paralysis, rape-induced, 240
partner rape, 184, 201–4
paternity uncertainty, 23, 24
Pathé, Michele, 162
patriarchal societies, 120, 121
patriarchy, 8, 120, 121, 175–76, 256–62, 265
PDAs (public displays of affection), 103

INDEX

perception biases, 42–45, 50–51, 171, 179, 250–54
persistence, sexual, 174–75
personality. *See also* Dark Triad
   and intimate partner violence, 139–40
   of mate guarders, 101–2
   and mate value, 74
   of rapists, 198–200
   of sexual exploiters, 52–53
   of sexual harassers, 180–81
   of women targeted by sexual exploiters, 46–47
personality disorders, 139. *See also specific disorders*
pets, 128, 150, 155
PhD programs, 173
physical abuse, 125
physical size, as level of power, 27
pickup-artist-training "boot camps," 14
Pinker, Steven, 205
Pisaura mirabilis, 9–10
plastic surgery, 269
plausible deniability, 110
Playboy, 64
"playing dead," 240–41
pleasure centers, 169–70
Plenty of Fish, 185
police, 122, 124, 164–65, 221
polygamy, 85, 190–91
polygyny, 18
polyspermy, 15
pornography, 52, 159–62, 180, 249, 270–71
possessiveness, 100, 263–67
post-traumatic stress disorder (PTSD), 74, 159, 244–47
potential mates, stalkers' threats against ex's, 156
power
   and gender equality/inequality, 88, 90, 120
   and intimate partner violence, 119, 141–42
   levers of, 26–28
   and mate preferences, 261–62
   and sexual assault, 191
   and sexual coercion, 204–5
   and sexual harassment, 175–76
   shifting the balance of, 265
pregnancy, 15–16, 131–32, 216–17

preparation hypothesis, 242
professions, female- vs. male-dominated, 173
Prosecco perception bias, 50–51
prostitution, 32, 34–36
proximity, power of, 27
psychological abuse, 122–28
psychopathy, 5, 29, 52, 56, 58–59, 83
PTSD. *See* post-traumatic stress disorder
public displays of affection (PDAs), 103

quid pro quo sexual harassment, 174–75, 258

rage, 153, 199, 200, 215, 253
rape, 181–204
   characteristics of men especially prone to, 198–201
   concealment of, 219–21
   costs of, for victims, 216–21
   educating men about horrors of, 253–54
   evolutionary perspective on, 180, 183–84, 186–93
   fear of, 182–85, 226–34
   in human history, 209–14
   marital/partner, 201–4, 256–58
   men's minimization of effects of, 214–15, 253
   physical resistance to attempts at, 238–39
   and sexual arousal, 188–89
   and specialized rape adaptation hypothesis, 186–93
   stranger vs. acquaintance, 181–86
   underreporting of, 243
   and universality criterion, 193–98
   women's actions aimed at avoiding, 225–26
rape-as-adaptation theory, 187–89
"rape culture," 208
rape-induced paralysis, 240
The Rape of Nanking (Chang), 211
Rape Warfare (Allen), 211
"reasonable person" standard, 254–56
recalibration theory of anger, 108
recklessness, 48, 54
refuging, in PTSD, 245
rejected lovers, stalking by, 150–57
"The Rejected Stalker," 152–57

INDEX

relationship conflict, 92–117
  adaptive responses to, 92
  becoming irreplaceable as response to, 114–17
  and jealousy, 93–98
  and mate guarding, 99–106
  and mate-value discrepancies, 107–11
  serial mating as response to, 111–14
relationship load, heavy, 150, 153
relationship satisfaction, 74
religious institutions, codification of rape in, 190
reproductive skew, 18
Republic of Congo, 173, 254
resisting control (tactic against mate guarding), 104–5
resource infidelity, 79–84
restraining orders, 119, 124, 162, 163, 165
revenge porn, 159–62
reward circuitry, 169–70, 204
risk avoidance, 237
risk taking, 48
robbery, 46, 192
Roberts, Tomi-Ann, 267
Rock, Chris, 35
Rohypnol, 26, 182, 276
Romania, 95
Rose, Charlie, 7–8
Russell, Diane, 215
Russia, 119
Rwandan civil war, 195, 217

Sages of the Talmud, 210
Santa Cruz, California, 271
Saudi Arabia, 23–24
scandals, sexual, 7–8
Scandinavia, 24, 212, 213
Scelza, Brooke, 95–97
Schmitt, David, 31
Schwarzenegger, Arnold, 35–36
Seabright, Paul, 30
Seattle, Washington, 237
Sebold, Alice, 239, 254
Second Congo War, 211
"second crime paradox," 231
secret bank accounts, 80
self-blame, in victims of sexual assault, 241–43, 247
self-deception, 40

self-defense classes, 239
self-esteem, 37, 124, 136, 217–19
self-objectification, 267–68
selfishness, 108
Sell, Aaron, 108
Semai, 210
serial harassers, 28–29
serial mating, 111–14
sex (biological)
  and gamete size, 4
  individual differences within each, 5
sex differences, 4–5
  and battle over women's bodies, 15–16
  in evaluation of attractiveness, 11–12
  in jealousy, 94–98
  in optimum timing for initiating sex, 13–14
  in perception of sexual coercion, 253–54
  in perception of sexual disgust, 251–53
  in stalking, 147
  in victims of harassment, 175
sex ratios, 206, 272–73
sexting, 252
sexual aggression, 214–15. *See also* sexual coercion
sexual arousal, 188–89, 204
sexual attention, unwanted, 11, 168–72
sexual bribery, 178
sexual coercion, 166–247
  bodyguard hypothesis and defense against, 222–26
  concealment in aftermath of, 243–44
  definition of, 166
  from evolutionary perspective, 166–68, 172–73, 180, 183–84, 215–16
  fear of, 226–34
  harassment, 172–81
  in human history, 209–14
  and men's mating psychology, 204–6
  post-traumatic stress symptoms in victims of, 244–47
  rape, 181–204, 214–22
  sex differences in perception of, 253–54
  and unwanted sexual attention, 168–72
  women's defenses against, 234–42
sexual conflict
  among Tiwi, 17–21
  coevolution of, 10
  complexity of, 8–9

INDEX

sexual conflict (*cont.*)
  costs of, 21–22
  manifestations of, 3
  on mating market, 30–31
  in the natural world, 9–11
  and need for cooperation, 12–15
  and perception biases, 42–43
  psychological toll of, 250
  and resource infidelity, 79–84
  social circumstances influencing, 5–6
sexual conflict theory, 10–11
sexual disgust, 251–53
sexual double standards, in mateships, 84–87
sexual fantasies and fantasizing, 89–90, 195–96
sexual favors, 59, 77, 89, 174, 176, 200, 258
sexual harassment, 172–81
sexual intercourse
  sex differences in optimum timing for initiating, 13–14
  what "counts" as, 85–86
sexual jealousy, 22–25, 92–98
Sexual Orientation Inventory (SOI), 105
sexual over-perception bias, 42–44, 171, 179, 251
sexual persistence, 174–75
sexual predators, 182–86, 189–92
  and concealment, 220, 221, 249
  and Dark Triad, 28–29
  guarding against, 51, 225, 245, 276
  on internet, 185, 186
  stalkers as, 151
sexual proprietariness, male, 264
sexual psychology, evolution of, 22–26
sexual reputation, revenge porn and, 160–61
sexual variety, desire for, 31–36, 84–85, 204
sexual withdrawal and bestowal, in mateships, 87–91, 108
shame
  in rape victims, 217
  in stalkers, 153–54
  in victims of intimate partner violence, 125–26
shared vision, 117
Sheen, Charlie, 36

short-term mating strategy. *See also* casual sex
  and Dark Triad traits, 53, 58, 82–83, 169, 180, 191, 200–201
  and mate poaching, 129
  and rape, 200–201
  and sexual coercion, 191
  and sexual harassment, 180
  and sexual over-perception bias, 43–44
  and status-enhancing displays, 82
  and visual cues, 169
  in women, 40, 53–54
Shriver, Maria, 36
Shuar, 96
shyness, 47
signaling, sexual, 45–46, 53–55, 81
Simpson, Nicole Brown, 149
Simpson, O.J., 149
slavery, 191, 265
smiling, 54–55, 76, 103, 109–10, 179, 251
Smith, Joseph, 84–85
Smith, Susan, 134
Smuts, Barbara, 216, 224
snakes, 227
Snider, Paul, 64, 75
social circumstances, affecting sexual conflict, 5–6
social distancing, 138
social isolation, in rape victims, 217–18
social learning theory, 120, 121
social media, stalking and exposure on, 164
social mind reading, deficits in, 151
social services workers, 173
social status, 189–92
social support, for victims of stalking, 163
socioeconomic status, 189–92
soft rejections, 179–80
SOI (Sexual Orientation Inventory), 105
South Africa, 191, 203–4
South America, 32
South Dakota, 212
South Korea, 95, 97
Spain, 95
speed dating, 43
spending habits, 80–81
sperm, 4, 15
spiders, 9–11, 181, 227
spousal rape, 201–4
spy cams, 23
Sri Lanka, 198

324

INDEX

stalking, 145–65
  defenses against, 162–65
  "effectiveness" of, 157–58
  evolutionary perspective on, 146–47
  examples of, 145
  and the law, 145–46
  and negative reinforcement contingency, 158
  physical/emotional toll on victims of, 148–50
  "reasonable person" standard, 255
  by rejected lovers, 150–57
  and revenge porn, 159–62
  sex differences in, 147
stalkinghelp.org, 162
Stanford University, 199, 243, 253, 271
status, perceived, 81, 82
stepchildren, 62, 133–35, 217, 228
stotting, 45–46
stranger fear, 185, 228, 232
strategic interference, 30–31, 62
Stratten, Dorothy, 64, 75
strength, as level of power, 27
submissive behavior, 101
suicide and suicidal thoughts, 155, 159, 162, 220
Surviving Stalking (Pathé), 162
Sweden, 13, 24, 95, 121, 170–71, 229
Symons, Donald, 61, 87, 90, 193
Sznycer, Daniel, 125

Tanzania, 96
technology
  and access to pornography, 270
  and mate guarding, 104
Ten Commandments, 197, 264–65
testosterone, 90
Texas, 252, 259
text messaging, 145
theft, 46
Theron, Charlize, 269
Thomson, Andy, 50
Thornhill, Randy, 187, 192
threats of future harms, 128, 137
threats of violence, 148, 155–56, 205
time of day, rape fear and, 230
Tinder, 12, 33, 140, 185, 272
Tiwi, 17–21, 184
Tokyo, Japan, 46, 138–39

tonic immobility, 239–41, 245
torso shape, 223
toxic masculinity, 7, 8
treachery, sexual, 38–41
Trump, Donald, 191
Tsimane, 96
tummy tucks, 269
Turkey, 31, 266
Turner, Brock, 199, 243, 253, 271
Tutsi, 195
Twitter, 164

Ullman, Tracey, 84
unadorned attractiveness, 82
under-perception biases, sexual, 44–45
United Arab Emirates, 256–57
United Kingdom, 170, 252, 254, 258, 274
United States
  intimate partner violence in, 119–20, 274
  marital rape in, 257
  sexual harassment complaints in, 175
  stalking laws in, 145–46
universality criterion (with rape), 193–98
University of California, Berkeley, 259
University of California, Santa Barbara, 76
University of Michigan, 259
University of Texas at Austin, 53–54, 255, 259, 272
unwanted sexual attention, 11, 168–72

V-shaped torso, 223
vaginal lubrication, 242
Valero, Helena, 254
variety, desire for sexual, 31–36, 84–85, 204
veiling, 141
Venezuela, 173
verbal abuse, 124
victim blaming, 241, 243–44
vigilance, 99–102, 234–35, 237
Vikings, 213
violence. *See also* intimate partner violence (IPV)
  mate-related, 98
  from rejected stalkers, 148–49
  threats of, 101, 205
Violence Against Women Act, 120
virginity tests, 23

INDEX

vision, shared, 117
visual stimuli, 78, 168, 204, 234, 253

Walsh, Wendy, 179
warfare, rape in, 209–11
watches, expensive, 39
water striders, 11
weapons, 142–43
Weinstein, Harvey, 7, 174, 181, 182, 189–90, 249
WEIRD cultures, 95–96
welfare trade-off ratios (WTRs), 74, 107–12, 126
Wells, Rebekah, 160, 162
Westermarck, Edvard, 190
Western cultures, 79–80, 88, 95–96
Whole Foods, 82
widow remarriage, 18
wife beating, laws against, 274
Wild (online dating site), 272
Wilson, Margo, 130, 137, 143, 216, 222–24, 264
withdrawing from threats, 237–38
witnesses of sexual harassment, 259
women. *See also* intimate partner violence (IPV); sex differences
  assault by, as defense against intimate partner violence, 142–43
  attachment-avoidant, 106
  "choosiness" of, 12
  on Dark Triad, 58–60, 78
  defenses of, against sexual coercion, 234–42
  as deflectors of unwanted sexual attention, 55
  and desire for sexual variety, 31–36
  empowerment of, 265–67
  evoking of jealousy by, 109–10
  exploitability cues in, 47–53
  fear of rape/murder in, 182–85
  infidelity among, 77
  infidelity as defined by, 86–87
  jealousy in, 95, 96
  marital status and sexual victimization of, 224
  mate guarding in, 101
  mate preferences of, 261–63
  and mate value, 75–76, 98
  resistance to mate guarding in, 102–6
  self-esteem in, 37
  sex differences between men and, 4–5
  sexual objectification of, 267–72
  as sexual predators, 29
  as stalking victims, 147–57
  submissive behavior in, 101
  in Tiwi culture, 17–21
  under-perception bias in, 44–45
Women Against Revenge Porn, 160
women's bodies, and reproduction, 15–16, 94
women's magazines, 39
Woodward, Joanne, 35
work environments, modern, 173–75, 258–59
World War II, 211
WTRs. *See* welfare trade-off ratios

Y chromosome, 213
Yanoáma (Valero), 254
Yanomamo, 27, 141, 173, 184, 210–11, 226
Yap, 211
Yasawa, 96
Yemen, 202
young age, as cue, 49–50

Zahavi, Amotz, 81
Zambia, 228
zero-tolerance policies, 249
Zimbabwe, 90
Zuckerberg, Mark, 82

ABOUT THE AUTHOR

DAVID M. BUSS is a professor of psychology at the University of Texas at Austin and a past president of the Human Behavior and Evolution Society. He is the author of several books, including *The Evolution of Desire*, *The Dangerous Passion*, *The Murderer Next Door*, and *Why Women Have Sex* (co-authored with Dr. Cindy Meston). He has written for publications such as the *New York Times*, the *Los Angeles Times*, and *Psychology Today*, and he has made more than thirty television appearances on shows including *CBS This Morning*, ABC's *20/20*, and NBC's *Dateline* and *Today*, among others. Buss has received numerous awards, which include the Distinguished Scientific Award for an Early Career Contribution to Psychology by the American Psychological Association (APA), a fellowship at the Center for Advanced Study in the Behavioral Sciences at Stanford, and the Association for Psychological Science (APS) Mentor Award for Lifetime Achievement (2017). He has been cited as one of the fifty most influential psychologists in the world.